LIFE
HACKS

Written by Julia March and Rosie Peet
Models by Barney Main and Nate Dias

CONTENTS

★ ★ ★ ★ ★ ★ ★ ★ ★ ★ ★ ★

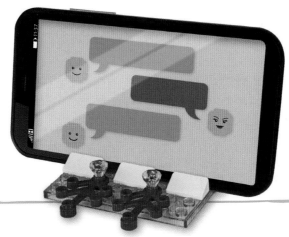

Weekly planner

Weighing scales

Playing card holder

Bookmarks

KEY HANGERS

★ ★ ★ ★ ★ ★ ★ ★

You'll never forget your key when it's hanging from an eye-catching LEGO® key hanger. You could make your hanger in the shape of a tree, or even a chameleon! Anything with a long "branch" that sticks out will work. As well as keys, you can use these hangers for jewellery or other trinkets.

How to build:

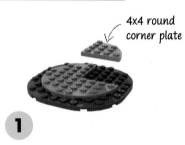

4x4 round corner plate

1

Start by building a round base for your hanger out of two layers of plates.

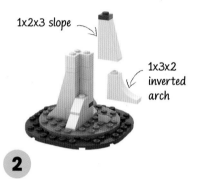

1x2x3 slope

1x3x2 inverted arch

2

Build a trunk using slopes and arches. Face the arches in different directions to give the trunk stability.

1x2/2x2 angle plate

Brown plates add colour and a sense of texture

3

Increase the height of the tree trunk using bricks and plates. Build in angle plates so that you can add branches.

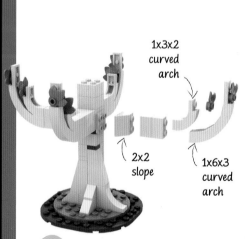

1x3x2 curved arch

2x2 slope

1x6x3 curved arch

4

Build branches onto the angle plates. Add curved arches to the ends of the branches. Decorate with leaf pieces.

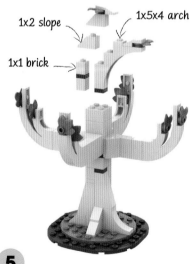

1x2 slope

1x5x4 arch

1x1 brick

5

Add a top branch, using a 1x5x4 arch and a stack of small bricks and slopes.

6

Add another top branch behind the first one, curving the opposite way to balance out the shape. Add leaves.

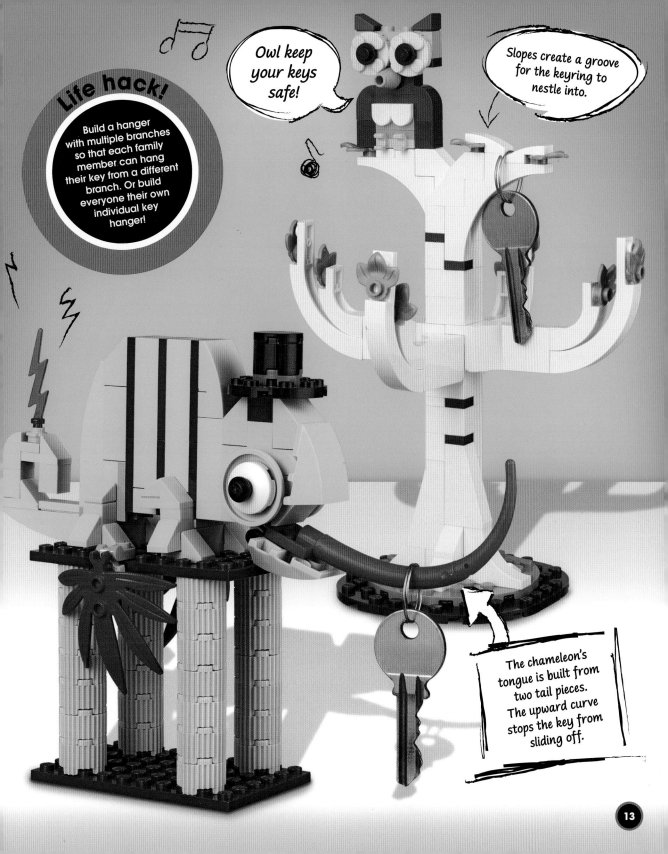

13

BOOKMARKS

★ ★ ★ ★ ★ ★ ★ ★ ★ ★ ★ ★ ★

Never lose your page again!

Next time you take a reading break, save your place the stylish way – with a LEGO bookmark. Build one with a minifigure topper, or experiment with a bold and beaky bird. It's much kinder to your books than folding over a page corner!

How to build:

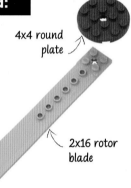

4x4 round plate

2x16 rotor blade

1

Place a 4x4 round plate onto a 2x16 rotor blade, letting the plate hang off the edge of the blade.

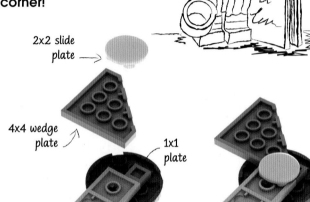

2x2 slide plate

4x4 wedge plate

1x1 plate

2

Add a 4x4 wedge plate and a 1x1 plate to the back of the rotor blade and secure with a 2x2 slide plate.

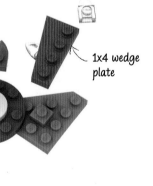

1x1 tooth plate

1x4 wedge plate

3

Bring your bird to life with a wedge plate for a wing, tooth plates for feathers, and an eye piece.

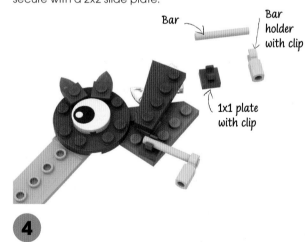

Bar

Bar holder with clip

1x1 plate with clip

4

Finally, add some angular legs made from yellow bars and bar holders.

Your bookmark can be any design. Just make sure to include long flat pieces. Lightning bolts work perfectly!

Decorate your bookmarks with animals, abstract patterns, or minifigures.

The bird's long beak piece fits perfectly between the pages.

CODE BREAKER

★ ★ ★ ★ ★ ★ ★ ★ ★ ★

The code fits into the code breaker to reveal the message: HEY!

It's fun to exchange secret messages with a friend. These cunning codes will keep your communications clandestine, whether they are casual chat or juicy pieces of top-secret news.

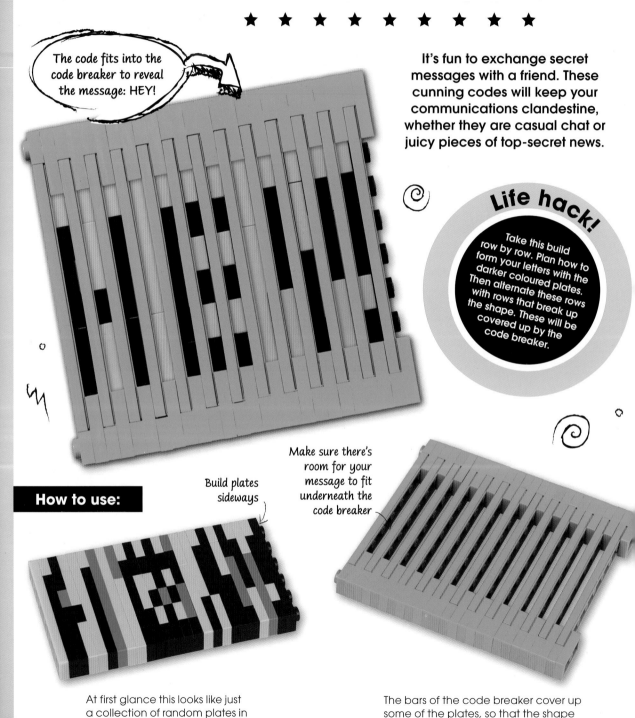

Life hack!

Take this build row by row. Plan how to form your letters with the darker coloured plates. Then alternate these rows with rows that break up the shape. These will be covered up by the code breaker.

How to use:

Build plates sideways

Make sure there's room for your message to fit underneath the code breaker

At first glance this looks like just a collection of random plates in different shades of blue.

The bars of the code breaker cover up some of the plates, so that the shape of the letters show through the gaps.

SECRET MESSAGE

★ ★ ★ ★ ★ ★ ★ ★ ★ ★

Yikes... I've got code overload!

In this code, a letter is represented by a pair of plates in the colours shown above and to the left of it on the grid. For example, an orange plate on top of a transparent blue plate is code for the letter "Y". Line up or stack your plate pairs to spell out words.

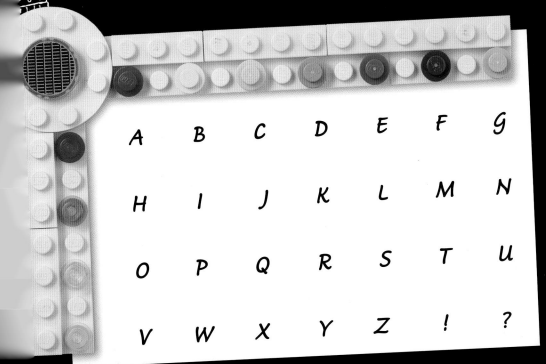

A	B	C	D	E	F	G
H	I	J	K	L	M	N
O	P	Q	R	S	T	U
V	W	X	Y	Z	!	?

use:

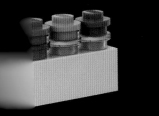

= H

= E

= Y

= !

= W

= O

= W

= !

tters next to each other in the order
uld be read. You could write your
kwards to make it harder to crack!

Alternatively, stack your letters on top of each
other. Choose whether your message should
be read from the bottom up or top down

PLAYING CARD HOLDERS

★ ★ ★ ★ ★ ★ ★ ★

Make sure the cards are stacked in your favour with this tasteful playing card holder. The two-tier holder will keep your cards visible to you, and only you, while you decide which one to play next. It's a winner – and hopefully you will be too.

I always keep a joker face!

How to build:

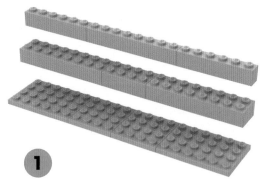

1

Start by lining up 4-stud-wide plates. Add a line of 2-stud-wide bricks, then a line of 1-stud-wide bricks.

2x4 curved slope

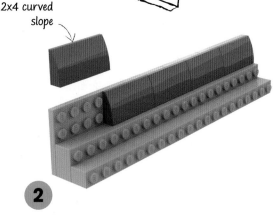

2

Turn the build on its side. Next, place 2x4 curved slopes along the top edge. Add another row below it.

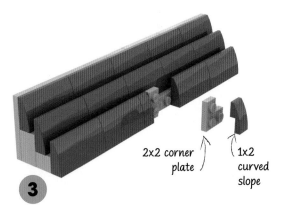

2x2 corner plate

1x2 curved slope

3

Add more slopes to the front edge. Leave a gap for two 2x2 corner plates and two 1x2 curved slopes.

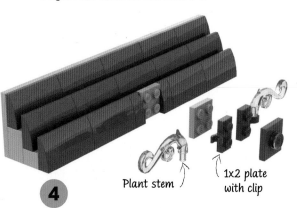

Plant stem

1x2 plate with clip

4

Add decorative flourishes such as plant stem pieces to the gap in the centre of your card holder.

Six LEGO playing cards fit perfectly between the grooves of this card holder.

This alternative design is built slightly differently, but still has a front, middle, and back edge.

DECISION MAKER

★ ★ ★ ★ ★ ★ ★

Are you going to tidy your room? Or are you going to leave it for another day? Let your LEGO build decide! Drop a ball into this decision maker and where it lands will give you your answer. Will it be yes, no, later, ask again, or shove everything under the bed and hope for the best?

Drop your LEGO ball piece through this hole at the top of the model.

But I just tidied it last year...

Minifigure heads with varied expressions represent the results.

16x16 plate

1x1 brick with side stud

1

Start with a 16x16 plate. Place two 1x1 bricks wih side studs at one end, spaced two studs apart.

2

Build the edges of your decision maker using two layers of bricks. Create five sections at one end.

4x4 round plate with hole

3

Complete the hole at the top by adding a 2x4 plate and a 4x4 round plate with hole.

One-stud-wide plates

1x2 brick with side studs

1x2 jumper plate

4

Create five square shapes using one-stud-wide plates. Add bricks with side studs and jumper plates.

1x1 round brick

5

Place stacks of 1x1 round bricks in an inverted v-shape. These will be the pins for the ball to bounce off.

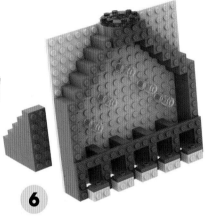

6

Build two supports and add them to the back of the baseplate so that your build can stand upright.

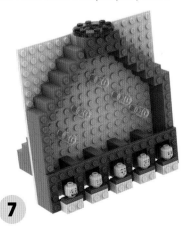

7

Add a minifigure head to each jumper plate at the bottom. Now all you need is a LEGO ball piece!

I can predict which decision you'll make...

Life hack!

This build could double as a game. Assign a score to each of the minifigure heads and take turns with a friend to drop the ball through the hole. See who can score the highest in five goes.

21

SPINNER

★ ★ ★ ★ ★ ★ ★ ★

Do you and your friends ever argue about what movie to watch or who should take the next turn in a game? Cut the stress and let this spinner make the choice for you. The person whose colour it lands on gets the final word.

Give this build a whirl!

How to build:

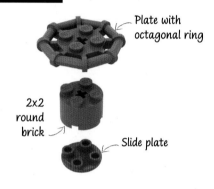

Plate with octagonal ring

2x2 round brick

Slide plate

1

Build the centre of the spinner from a slide plate, a round brick, and a plate with octagonal ring.

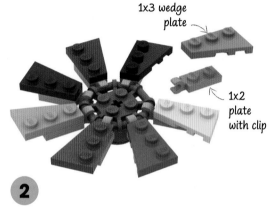

1x3 wedge plate

1x2 plate with clip

2

Connect 1x2 plates with clips around each side of the ring. Add 1x3 wedge plates in different colours.

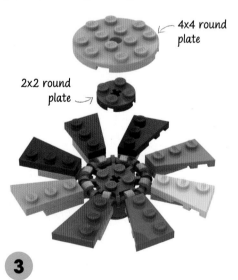

4x4 round plate

2x2 round plate

3

Add a 2x2 and a 4x4 round plate onto the plate with octagonal ring.

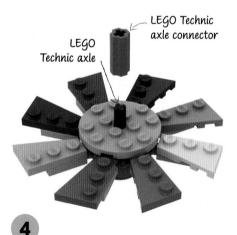

LEGO Technic axle connector

LEGO Technic axle

4

Finally, insert a LEGO® Technic axle through the centre of the round plates. Place an axle connector over it.

Make sure your plates are the same length so that the spinner is weighted evenly.

Life hack!

Instead of picking a colour and seeing who wins, you could assign an outcome to each colour. With this build, there are eight possible outcomes!

The spinner will come to rest on just one colour.

WEEKLY PLANNER

★ ★ ★ ★ ★ ★ ★ ★ ★

Build a family calendar and never forget a swimming lesson, art club, or music practice again. Choose a different minifigure for each family member. You don't want to get mixed up and end up at the salon having your dad's haircut!

Create numbers using plates for each day of the week.

My schedule is packed!

Your minifigures sit on the angle plates. Connect the backs of their legs to the studs of the baseplate.

16x16 plate

1x2/2x2 angle plate

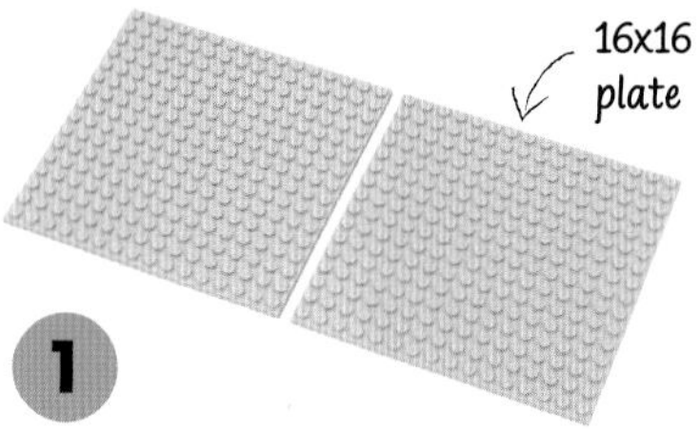

1
Give yourself lots of space to lay out your planner. Place two 16x16 plates side by side.

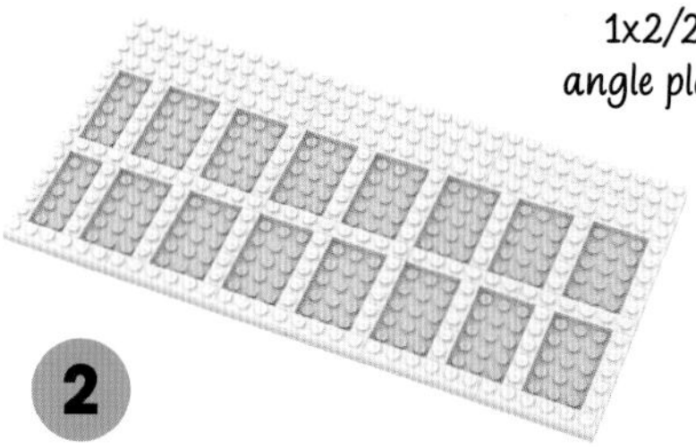

2
Use white plates to create a grid with eight columns. The first column will be narrower than the others.

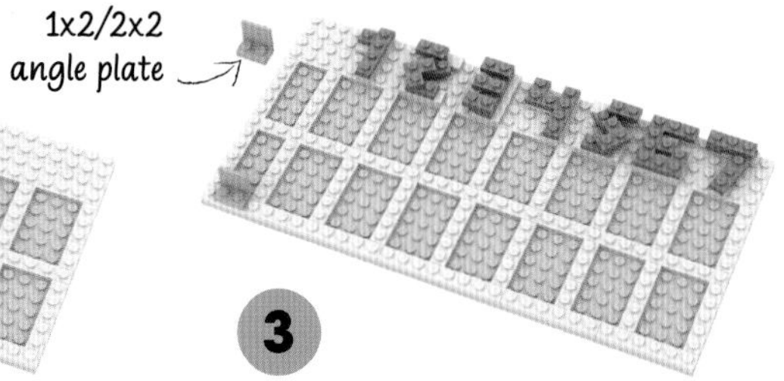

3
Add angle plates for your minifigures. Now add small builds to the squares to show your weekly activities!

Rummage through your LEGO collection for useful elements that can represent activities.

Plates with clips are useful for connecting small elements.

DECORATIVE BAG CLIPS

★ ★ ★ ★ ★

Help a friend resist the urge to open a gift bag before the big day with a LEGO bag clip. If you want to tantalize them, use the clip to hint at the gift. Do the banana pieces mean there are banana-flavoured sweets inside?

What could it be?

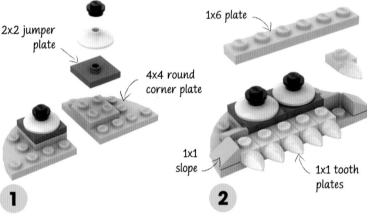

2x2 jumper plate

4x4 round corner plate

1x6 plate

1x1 slope

1x1 tooth plates

1x2 plate with ball

1x2 plate with socket

1x3 curved slope

1

Start by building a monster face onto two round corner plates. The eyes are secured with jumper plates.

2

Add a row of tooth plates, with slopes either side. Place a plate on top, so that only the tips of the teeth are visible.

3

Connect a plate with a ball to a plate with a socket. Place two plates and a 1x3 curved slope on top.

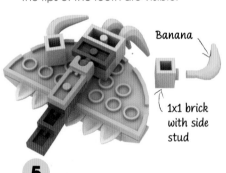

Banana

1x1 brick with side stud

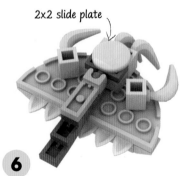

2x2 slide plate

4

Connect the two sections as shown. The ball and socket allow the clip to open and close over your bag.

5

Finish with some decorative touches. Banana pieces connect to bricks to form a punky monster hairdo.

6

Secure the plates at the back with a 2x2 slide plate. Turn your build over and your monster bag clip is ready to go!

If your clip is securing a gift for a friend, why not make it in their favourite colour?

Life hack!

These clips add a personal touch to paper gift bags – and discourage any peeking!

WEIGHING SCALES

★ ★ ★ ★ ★ ★ ★

Ever wondered how heavy your LEGO builds or minifigures are? Or whether a bunch of LEGO bananas weighs more than a few LEGO apples? These clever weighing scales let you compare the weights of small builds, minifigures, and other elements.

These scales are worth the weight!

If the two objects weigh roughly the same, the scales will balance and this part will stay level.

They say life's all about balance...

Place whatever you would like to weigh on the dish.

Life hack!

You could build your own LEGO weights using a 2x2 brick as your unit of measurement. Stack the bricks together to make the weights. How many bricks does your eraser or pen weigh?

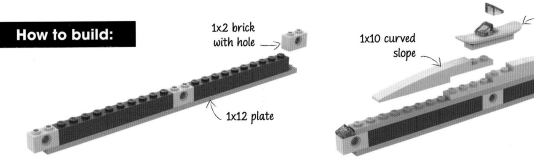

1x2 brick with hole

1x12 plate

1x10 curved slope

Minifigure snowboard

1 Add one-stud-wide bricks to two 1x12 plates. Place a 1x2 brick with a hole in the centre, and one at either end.

2 Create a rounded top with plates and curved slopes. Place some decorative pieces in the centre.

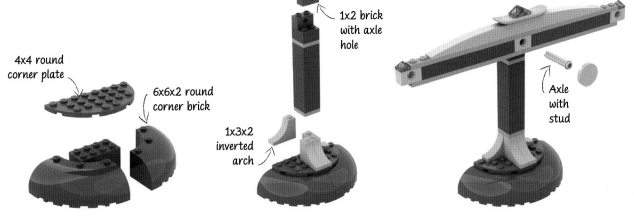

4x4 round corner plate

6x6x2 round corner brick

1x3x2 inverted arch

1x2 brick with axle hole

Axle with stud

3 Make the base out of two round corner bricks and two 2x4 bricks, secured by two round corner plates.

4 Place four inverted arches together and add a column topped with bricks with axle holes.

5 Add the two sections together and secure with an axle with a stud. Finish with a round tile.

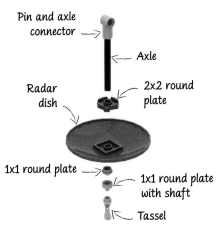

Pin and axle connector

Axle

Radar dish

2x2 round plate

1x1 round plate

1x1 round plate with shaft

Tassel

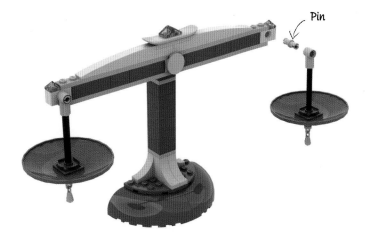

Pin

6 Connect a radar dish and round plate with an axle. Add a pin and axle connector to the top. Add decorative plates to the radar dish.

7 Connect the dish to one of the bricks with holes using a pin. Make another dish for the other side of the scales.

KEYRING

★ ★ ★ ★ ★ ★ ★ ★ ★

Do you find yourself searching for your keys every morning? Finding them can feel like searching for a needle in a haystack. Attach them to this bright tractor keyring and next time you'll spot them in a jiffy!

Oo arr! I'm locked out of my barn!

The keyring fits through a curved plate with a hole.

Use larger wheel pieces for the tractor's rear wheels.

How to build:

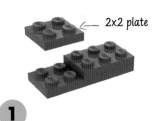

← 2x2 plate

1

Start with two layers of plates. Make the structure five studs long and two studs wide.

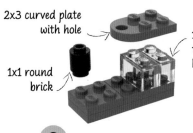

2x3 curved plate with hole

1x1 round brick

1x2 transparent brick

2

Add two 1x2 transparent bricks at one end, then a 2x3 curved plate with hole. Add a 1x1 round brick at the other end.

3

Add a third layer as shown, with a 2x3 curved plate with hole. Make sure the hole sticks out from the rest of the build.

A colourful rooster tops off this barnyard scene.

This key has come home to roost!

Life hack!

Once you've built your keyring, build somewhere to keep it safe and sound. The tractor rolls into this barn after a long day, ready for when you next need to find it!

Make your key's home wide enough to contain the key neatly.

1x2 plate with handle

4

Add a 1x2 plate with handle, with the handle sticking out, plus a 2x2 plate next to it.

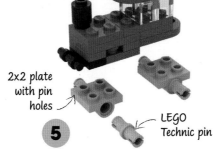

2x2 plate with pin holes

LEGO Technic pin

5

2x2 plates with pin holes form the next layer. Place LEGO Technic pins in the holes for the wheels to attach to.

6

Finally, place four wheel pieces onto the pins. Use small wheels at the front and larger ones at the back.

This trick is all about distraction! The "fake" lid is ornate and colourful, but the top and bottom of the secret compartment are made up of identical purple plates.

Hide your note in the secret compartment in the base.

Anyone who lifts the top lid off will just see an empty compartment.

NOTE BOX

This is a build to take note of!

★ ★ ★ ★

This little box looks like a simple container for trinkets, sweets, or paper clips. Only you and a trusted friend will know about the secret compartment in the base where you leave messages for each other. It's perfect for planning a prank, a party, or a birthday surprise!

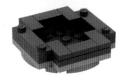

How to build:

6x6 round plate

1x2 plate

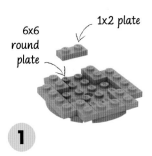

1

Take a 6x6 round plate and place 1x2 plates around the edges to form a cross shape.

1x4 plate with two studs

1x2 tile

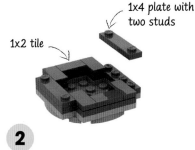

2

Add another layer of plates around the edge, then a layer of tiles and two 1x4 plates with two studs.

3

Cover with a layer of plates, then add bricks around the edge. Top with more tiles and plates with two studs.

4

Add two more layers of plates in contrasting colours for the lid.

2x2 round corner tile

4x4 round brick

5

Create a handle for the lid using another 6x6 round plate. Add a 4x4 round brick and 2x2 round corner tiles.

$2x2x1^2/_3$ dome

6

Add decorative touches such as yellow transparent slopes, a jewel connected by a flower piece, and a dome.

BRICK SEPARATOR DISPENSER

★ ★ ★ ★ ★ ★ ★ ★ ★ ★ ★

Inspirational build!

Pressing this red button forces the slopes below it downwards so they push open the door.

It can be tricky to break apart some bricks. That's where a LEGO brick separator comes in handy. Make sure you have one ready safely stored in this dispenser. When you need to demolish a build, just hit that button!

Life hack!

You could use this idea to store other useful objects that you'd like to have readily to hand.

Brick separator sits on a plate which tips forwards when the door is open, for easy access.

The door connects via pins and bricks with holes. The pins act as hinges that allow the door to open.

CHAPTER 2
★ ★ ★ ★ ★ ★ ★
TECH HACKS

Become a super hacker!

Tech tidy

Tablet stand

Headphone wrap

Phone stand

Phone speaker

PHONE STAND

★ ★ ★ ★ ★ ★

Hold, please!

Put your phone on hold with this decorative mobile phone holder. It will leave your hands free to do other stuff while talking, such as adding to your latest LEGO® creation! It can also be used to hold the phone securely while it is charging.

How to build:

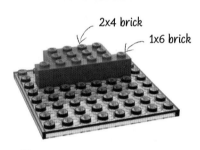

2x4 brick
1x6 brick

1

To make this ocean themed holder, start with an 8x8 transparent blue baseplate. Add 2x4 and 1x6 grey bricks.

1x1 plate with clip

2

Place plates with clips on either end of the 2x4 grey brick. The clips will hold your phone's cable when it is charging.

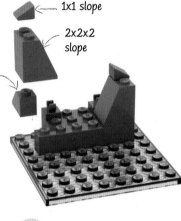

1x1 slope
2x2x2 slope
1x2 slope

3

Add a 1x2 slope, a 2x2x2 slope, and a 1x1 slope, all facing in different directions. Repeat on the other side.

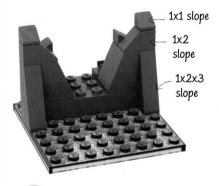

1x1 slope
1x2 slope
1x2x3 slope

4

Add a 1x2x3 slope topped with a 1x2 slope and a 1x1 slope. Repeat at the opposite end.

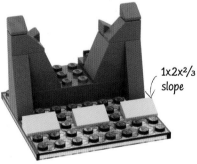

1x2x²/₃ slope

5

Place three light blue slopes in front of the grey bricks. Leave a gap one stud wide for your phone to sit in.

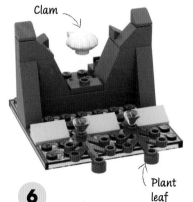

Clam

Plant leaf

6

Add decorative pieces to complete the ocean theme, such as a clam and pink leaf pieces topped with jewels.

Phone rests in between two rows of sloped bricks.

Flat edge of slope fits snugly against phone.

Clusters of grey sloped bricks look like rocky seabed.

Life hack!

Create your phone holder around any theme that appeals. If you like space, build one shaped like an asteroid! If you like animals, decorate your holder with cute minifigure pets!

HEADPHONE WRAPS

★ ★ ★ ★ ★ ★ ★

You get an urge to listen to a favourite song, you reach for your headphones, and... there they are, all twisted, tangled, and trailing. A neat wrap will ensure that in future, your headphones will be clean, tidy, and ready to bring sweet music to your ears.

I'm a big fan of wrap.

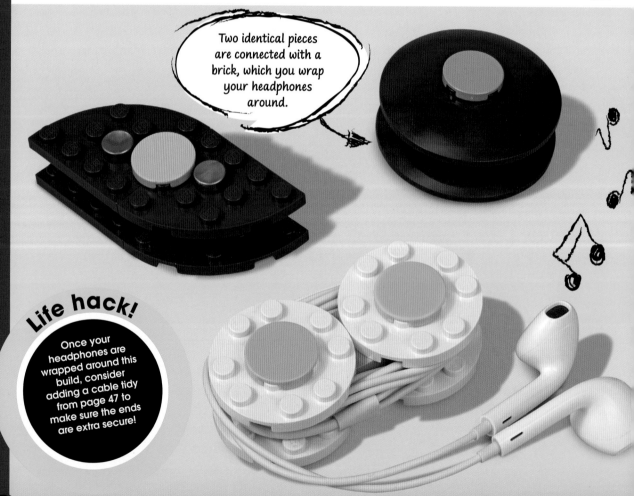

Two identical pieces are connected with a brick, which you wrap your headphones around.

Life hack!

Once your headphones are wrapped around this build, consider adding a cable tidy from page 47 to make sure the ends are extra secure!

How to build:

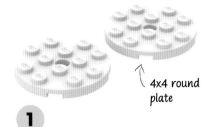

4x4 round plate

1

Start with two 4x4 round plates. These ones are white but they can be any colour you like.

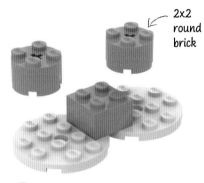

2x2 round brick

2

Connect the plates using a 2x2 brick, then add a 2x2 round brick on either side.

3

Add two more round plates directly over the first two. Place a 2x2 round tile in the centre of each plate.

CABLE TIES

★ ★ ★ ★

Loose cables don't just look messy, they can be hazardous, too. What if you tripped over one and knocked into your just-completed LEGO masterpiece? Hold those unruly cables together with these fun LEGO cable ties.

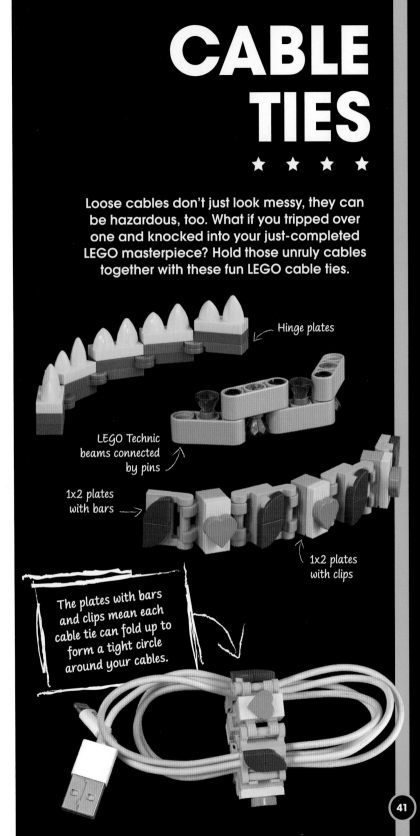

Hinge plates

LEGO Technic beams connected by pins

1x2 plates with bars

1x2 plates with clips

The plates with bars and clips mean each cable tie can fold up to form a tight circle around your cables.

PHONE SPEAKER

★ ★ ★ ★ ★ ★ ★ ★

This speaker rocks!

Here's a loudmouth you won't get tired of listening to! Start playing music on your mobile, slip it into the phone speaker's mouth, and hey presto – instant amplification! Why not decorate your speaker according to your favourite kind of music, too?

How to build:

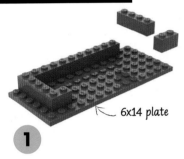

6x14 plate

1

Place one-stud-wide bricks on a 6x14 baseplate. Leave a gap eight studs wide at the front.

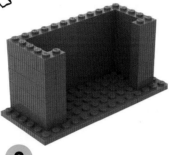

2

Continue building up in the same shape until you have five layers of bricks.

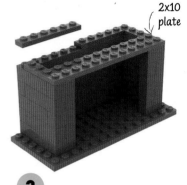

2x10 plate

3

Add a 2x10 plate across the front edge. Then add one-stud-wide plates along the remaining edges.

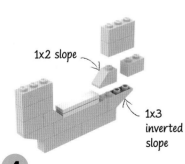

1x2 slope

1x3 inverted slope

4

Start building a minifigure face, leaving a gap for the mouth. Use a white plate and tile for teeth.

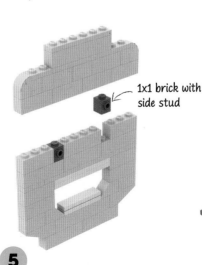

1x1 brick with side stud

5

Continue building up the minifigure face. Include two 1x1 bricks with side studs where the eyes will go.

2x2 round tile

6

Place the face onto the baseplate, covering the gap. Add two round tiles for the eyes.

TABLET STAND

★ ★ ★ ★ ★ ★ ★

Hands up if you want to watch a movie hands free! This stand is a secure place to lean your tablet, and places it at a comfortable angle for optimum viewing. Now just sit back and crack open the popcorn...

I hope this cinema has VIP seats...

How to build:

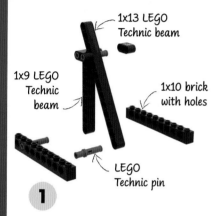

1x13 LEGO Technic beam

1x9 LEGO Technic beam

1x10 brick with holes

LEGO Technic pin

1

Connect a 1x9 and a 1x13 LEGO® Technic beam with pins and two 1x2 beams. Connect two 1x10 bricks with holes.

2

Repeat the previous step to make another identical structure.

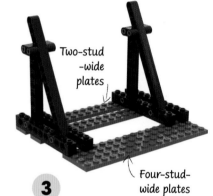

Two-stud-wide plates

Four-stud-wide plates

3

Place the two sections onto two rows of two-stud-wide plates, leaving a gap. Then add two four-stud-wide plates.

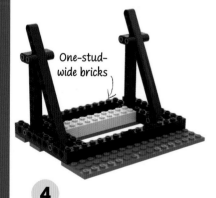

One-stud-wide bricks

4

Add a row of one-stud-wide bricks to the front and back plates. Add two-stud-wide bricks to the middle plate.

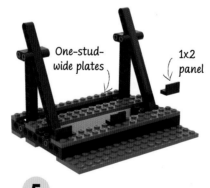

One-stud-wide plates

1x2 panel

5

Next, add plates to the front, middle, and back rows. Space 1x2 panels evenly along the front row.

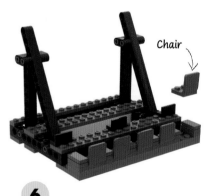

Chair

6

Finally, create a row of seats using chair pieces for your minifigures. Place 1x2 tiles in between for armrests.

Panels stop your tablet from slipping.

Line up your minifigures so they have a front-row view.

I hope the trailers aren't too long!

Life hack!

If you need your tablet to sit on a raised surface, consider building the base a few bricks higher.

TECH TIDY

★ ★ ★ ★

We all have stray bits and pieces relating to our tech – earbuds, USB sticks, and that miniature key that opens the sim tray on a mobile phone. It can really pile up! Keep it all organized with this storage box. It's ideal for tiny items that would get lost in a larger box.

Anyone seen my headphones?

Lid connects securely to these 1x1 plates on each corner.

Life hack!

If you'd rather have somewhere to store larger items such as remotes, tablets, or spare phone cases, build the model a little larger with just one section.

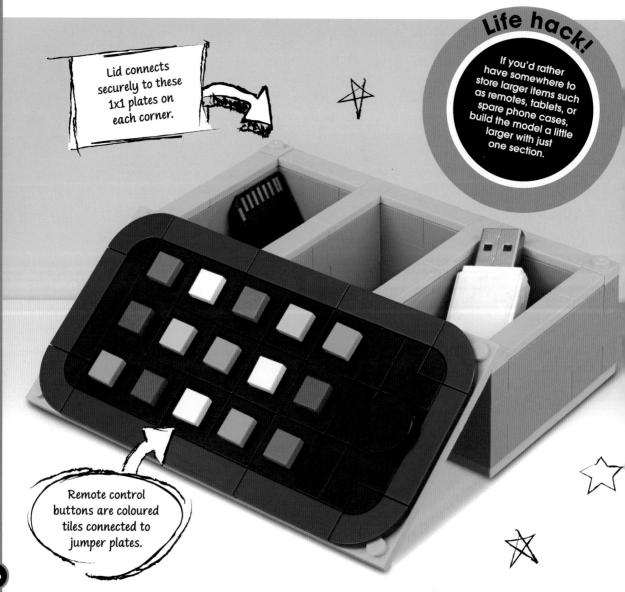

Remote control buttons are coloured tiles connected to jumper plates.

How to build:

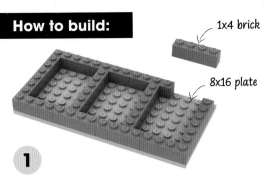

1x4 brick

8x16 plate

1

Start with an 8x16 baseplate and place a border of bricks around the edge. Divide into three rectangular sections using more bricks.

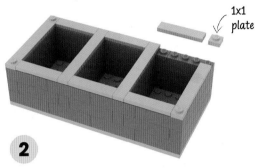

1x1 plate

2

Build up your box using two more layers of bricks. Add a layer of tiles to the top, with 1x1 plates in each corner for the lid.

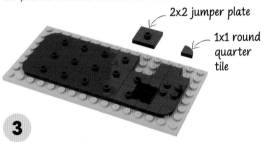

2x2 jumper plate

1x1 round quarter tile

3

Next make the lid. Take another 8x16 baseplate and add black tiles and fifteen jumper plates to attach "buttons".

2x2 round corner tile

4

Add coloured tiles and any other decor you like. Place tiles around the edge and round corner tiles in the corners.

CABLE TIDY

★ ★ ★ ★

It's easy to get your chargers and cables in a tangle – or lose them completely. Let your minifigures lend a helping hand! A phone charger fits perfectly into a minifigure hand, so you can avoid getting your wires crossed.

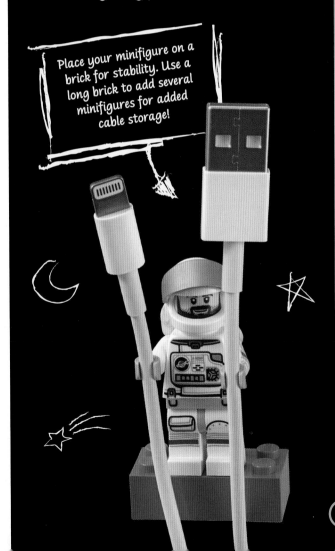

Place your minifigure on a brick for stability. Use a long brick to add several minifigures for added cable storage!

CHAPTER 3
★ ★ ★ ★ ★ ★ ★ ★
UPGRADE YOUR ROOM

Photo
frame

Torch

Tissue box
holder

Potted
plants

Create a cosy
space that's
all yours!

Bookends

POTTED PLANTS

★ ★ ★ ★ ★ ★ ★ ★

Bring some flower power into your life with these LEGO® potted plants – they never need watering and they are completely pest proof. Each plant is detachable from the pot, so if you're feeling a little prickly one day, just swap your petunia for a cactus.

Build yourself some flower power!

How to build:

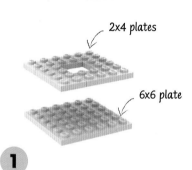

2x4 plates

6x6 plate

1

Start with a 6x6 plate. Add four 2x4 plates around the edge, leaving a square gap in the middle.

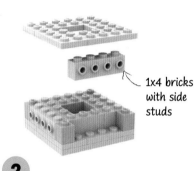

1x4 bricks with side studs

2

Add a layer of bricks, including bricks with side studs facing outwards, on all four sides. Add another layer of plates on top.

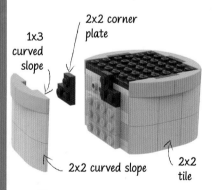

3

Add one more layer of plates, then repeat step two. Top with a brown 6x6 plate for soil.

2x2 corner plate

1x3 curved slope

2x2 curved slope

2x2 tile

4

Attach plates to the side studs. Create the pot's round edges using curved slopes and tiles and connect to the plates.

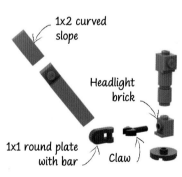

1x2 curved slope

Headlight brick

1x1 round plate with bar

Claw

5

Build a stem with round and headlight bricks. Attach a leaf to a headlight brick using a claw and a plate with bar.

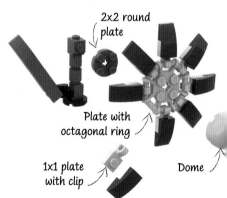

2x2 round plate

Plate with octagonal ring

1x1 plate with clip

Dome

6

Add a round plate, then a plate with octagonal ring. Add petals using plates with clips and curved slopes. Finish with a dome.

Spiky plant pieces connect to plates to form cactus spikes.

Detach the "soil" from the pot to swap the flower for another plant.

Choose bricks that match your room's colour scheme.

Life hack!

If you want your plants to "grow", add another brick to the stem every couple of weeks to increase the height. You'll have a thriving LEGO garden in no time!

MESSAGE BOARD

Time to check my messages!

★ ★ ★ ★ ★ ★ ★ ★

Leave little notes for friends and family on this LEGO message board. You don't have to use words – pictures work just as well. A broken heart could mean you're missing a friend, and a pizza shows what you'd like for lunch.

How to build:

2x4 brick

1x4 brick

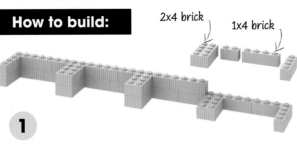

1
Build a wall with four sections.

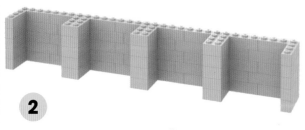

2
Continue building up your wall.

Headlight brick

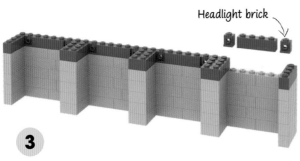

3
Add a final layer of bricks. Include two headlight bricks per section.

1x6x2 arch

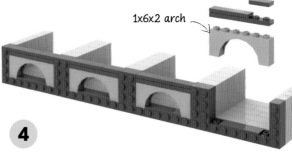

4
Turn the structure on its side. Connect arches to the headlight bricks. Place plates and tiles on top of the arches.

32x32 baseplate

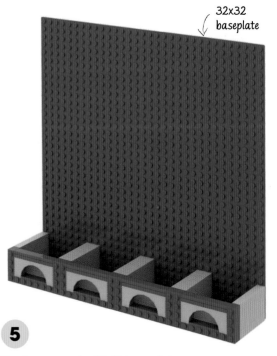

5
Finally, connect a 32x32 baseplate.

Add brightly coloured tiles around the edge of the baseplate for a decorative border.

Store LEGO bricks in these sections so that you have them on hand to create messages.

MONEY BOX

★ ★ ★ ★

Almost as good as a pot of gold!

This LEGO money box is more stylish than a traditional piggy bank, and after you break it open to get at your money it can be put together again. The size will depend on how much money you plan to save!

How to build:

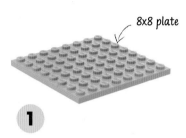

8x8 plate

1

Start with an 8x8 plate.

1x4 tile

1x2 tile

2x2 corner plate

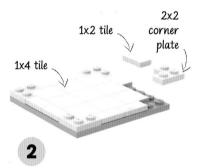

2

Add tiles over the top of the baseplate. Place 2x2 corner plates in each corner.

1x4 plate

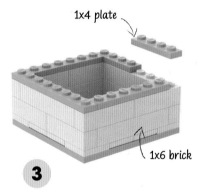

1x6 brick

3

Build up the box using bricks and a layer of blue plates to give a striped effect.

1x4 tile

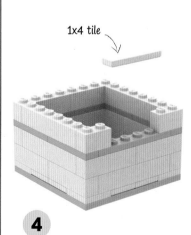

4

Add a further layer of bricks, leaving a four-stud-wide gap on one side. In this gap place a 1x4 tile.

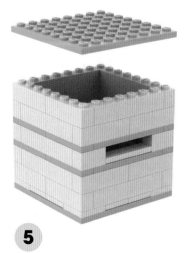

5

Continue building up the box using plates and bricks. When it is the desired height, place another 8x8 plate on top.

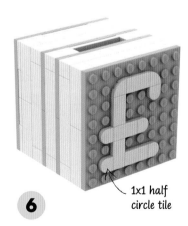

1x1 half circle tile

6

Finish by decorating your money box using tiles. When you've finished, turn the box over so the slot is at the top.

Spare change fits through the slot at the top.

Life hack!

Keep your money even safer by disguising the box! Add different decorations or patterns to the top that don't hint at your savings inside.

You can top your money box with whatever currency symbol you like.

55

Life hack!

The quarter planet design means there are a lot of bricks clustered together, making each bookend sturdy enough to lean a row of books against.

This build is easy if you plan-et!

Place minifigures that go with your theme along the top.

Build another bookend with the planet facing in the opposite direction. Place your books between the bookends.

BOOKENDS

★ ★ ★ ★ ★ ★ ★ ★ ★ ★ ★ ★ ★ ★

Here's a way to organize your space...

These bookends will keep your books standing neatly in line. Base them on scenes, characters, or a particular theme you like. How about some lunar landscape bookends to keep your space books from moonwalking off the end of the shelf?

How to build:

1

Place a large baseplate on top of a smaller one, with two-stud-wide plates around the edges.

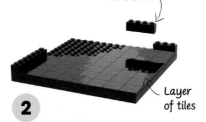

1x4 brick

Layer of tiles

2

Cover the large plate in tiles, leaving an area in the corner. Put one-stud-wide bricks at the edges.

3

Add yellow bricks to the corner. Continue to build up in gradually smaller layers to make a quarter sphere shape.

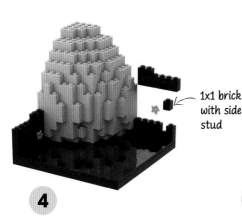

1x1 brick with side stud

4

Build up the sides with black bricks. Build in 1x1 bricks with side studs and connect stars to these.

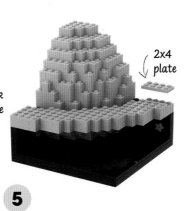

2x4 plate

5

Create a ring around the planet out of four layers of plates.

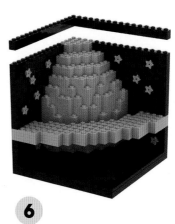

6

Add more black bricks, bricks with side studs, and stars to build up the two side walls.

DISPLAY SHELF

★ ★ ★ ★ ★ ★

Take your builds to new heights with a multi-level display stand. You could give your favourite build pride of place on the top level, and then use the other layer to display your favourite minifigures.

Express your-shelf!

It's tough at the top...

Shelves are wide enough to show off minifigures or create small scenes.

Wide "feet" give your display shelf height and add stability.

58

How to build:

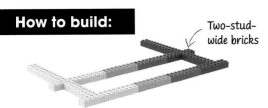

Two-stud-wide bricks

1 Build a frame out of two-stud-wide bricks.

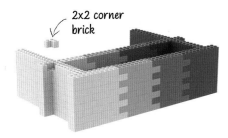

2x2 corner brick

2 Build up by another seven layers. Use corner bricks to secure the bricks going across to the bricks going down.

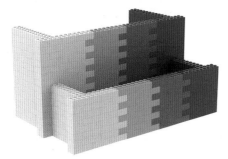

3 Build the back section up by another seven layers of bricks.

4 Turn your build over so that you have a wide lower shelf and a narrower top shelf.

"LIGHT-UP" LETTER

★ ★ ★ ★ ★ ★

You may never see your name in lights, but if you make this LEGO letter you can see lights in your name – or at least in your initial. The secret is the transparent bricks. Place the letter in front of a lamp or window and it will shine away as if it's all lit up.

Add transparent bricks at evenly spaced intervals for light to shine through.

DOOR HANGER

★ ★ ★ ★ ★ ★ ★ ★

It's annoying when someone knocks on your bedroom door when you're studying, gaming, or sleeping. This LEGO door hanger will tell people if it's OK to disturb you. Try a smiley face or thumbs up for yes, a slash symbol for no, or a couple of 'z's to say you're out for the count!

I need my eight hours, OK?

How to build:

1x2 brick with studs on two sides

1x1 brick with side stud

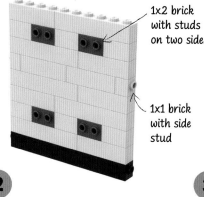

1x2 brick with studs on two sides

1x1 brick with side stud

1x5x4 inverted arch

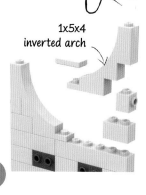

1

Build three rows of bricks in the width of your hanger. Include bricks with side studs facing to the sides and the front.

2

Build up your hanger. Include four bricks with studs on two sides, so you can add your designs to the front and back.

3

Add two 1x5x4 inverted arches to create a circular shape. Add another pair of bricks with sideways studs at either side.

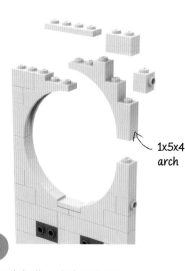

1x5x4 arch

1x2 curved slope

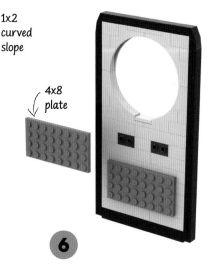

4x8 plate

4

Complete the circle using two more arch pieces. Square off the edges, adding in another pair of bricks with side studs.

5

Connect plates to the sides and top of your build, and then add tiles. Curved slope pieces create rounded corners.

6

Add two 4x8 plates to the front and then do the same on the back. This is the base where you can add your messages.

Add opposite messages on each side, for example a thumbs up on one side and a thumbs down on the other. Just flip the sign over depending on your mood!

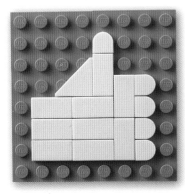

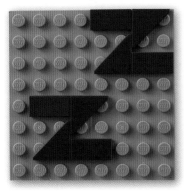

z Z z Z Z

The sign hangs over your door handle by this gap.

Keep your signs simple and easy to understand. This symbol clearly tells people to "keep out"!

TISSUE BOX HOLDER

★ ★ ★ ★ ★ ★ ★ ★

Don't make an issue out of finding your tissues! Keep them readily to hand in this LEGO tissue box holder. It can be fun to add a character – for example, a chef. The fluffy white tissues could double as his tall hat!

My compliments to the chef!

Before you start, measure around the tissue box with your bricks to make sure it will fit.

Add fun details to your build, like a frying pan on fire and a spatula.

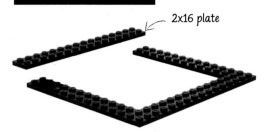

2x16 plate

1 Lay out four 2x16 plates in a square shape.

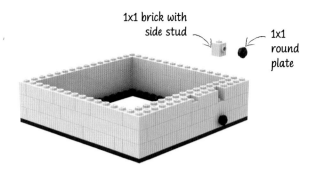

1x1 brick with side stud

1x1 round plate

2 Build up the box in a square shape, leaving a one-stud-wide edge free on the baseplates. Include bricks with side studs and round plates for buttons.

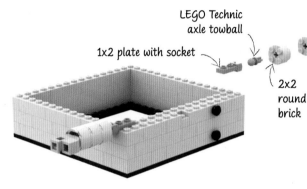

LEGO Technic axle towball

1x2 plate with socket

2x2 round brick

3 Add arms made from a plate with a socket, a LEGO® Technic axle towball, round bricks, and tan pieces for hands.

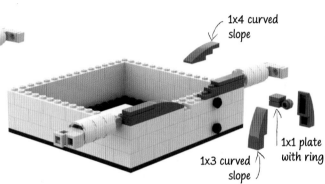

1x4 curved slope

1x1 plate with ring

1x3 curved slope

4 Add a neckerchief made from plates and curved slopes, connected by a plate with ring.

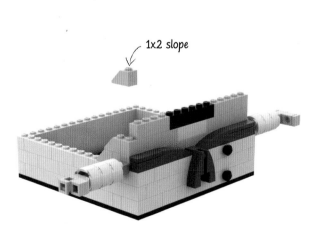

1x2 slope

5 Start building the chef's face. Include some slopes on each side wall.

Radar dish

6 Finish building the face up to the brim of the chef's hat. Include bricks with side studs to connect features like eyes, a nose, and a moustache.

Your hat rests on the top of the cactus.

The side arms help hold the hat in place, or can hold other small accessories.

Life hack!

Make sure your hatstand is tall enough, so that your hat doesn't touch the base. If it's not quite tall enough, just add more bricks to the stem.

Add finishing touches to the base, such as 1x1 round plates for stones, or a scorpion.

HATSTAND

★ ★ ★ ★ ★ ★ ★ ★ ★ ★ ★ ★

If you want to get ahead, get a hatstand. This cactus-shaped LEGO version is the perfect place to store your beanie hat and keep it looking sharp. The arms can be used to hold other trinkets, too.

Keep your hats looking sharp!

How to build:

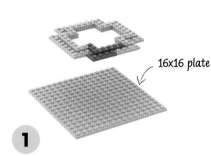

16x16 plate

1

Use tan plates to make the outline of a cross and attach them to a baseplate. This will form the shape of the cactus.

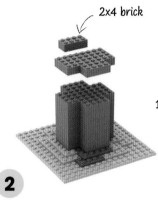

2x4 brick

2

Fill the cross-shaped border with 2x4 bricks. Stack upwards until the stem reaches eight bricks in height.

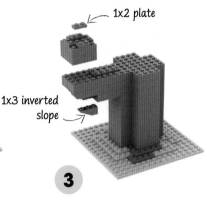

1x2 plate

1x3 inverted slope

3

Now add an arm. Inverted slopes give it a bendy look. Use more bricks and plates to complete the shape.

4

Add another arm on the opposite side using inverted slopes, bricks, and plates.

5

Keep stacking up the stem with 2x4 bricks in the cross-shaped pattern. Build up as many layers as you like.

5x5 round corner brick

6

Form a circle, using 5x5 round corner bricks at the top of the stem. This will help your hat stay in place on the stand.

TORCH

★ ★ ★ ★ ★ ★ ★

If you're full of bright ideas, this is the project for you. It's a fully functioning torch – just what you need when you're searching under the bed for that pesky LEGO piece you've lost. A LEGO light brick provides the illumination you need at the flick of a switch.

Call that a torch?

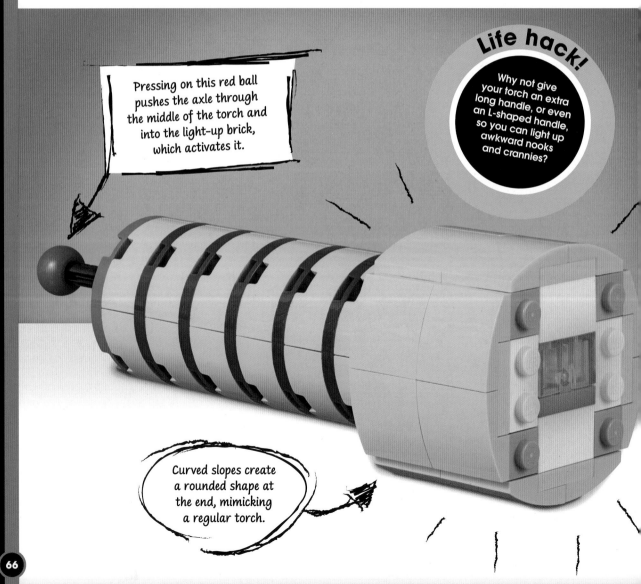

Pressing on this red ball pushes the axle through the middle of the torch and into the light-up brick, which activates it.

Life hack!

Why not give your torch an extra long handle, or even an L-shaped handle, so you can light up awkward nooks and crannies?

Curved slopes create a rounded shape at the end, mimicking a regular torch.

How to build:

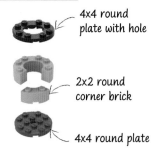

- 4x4 round plate with hole
- 2x2 round corner brick
- 4x4 round plate

1

Place four round corner bricks in a circle on a 4x4 round plate. Add a 4x4 round plate with hole on top.

2

Build five more layers of corner bricks and round plates with holes, so that you have a long tube.

- LEGO Technic bush
- Axle connector
- Axle
- LEGO Technic ball joint

3

Connect two axles and place a LEGO Technic bush on top. Put this through the centre of the tube and the hole in the plate. Add a ball joint to the end.

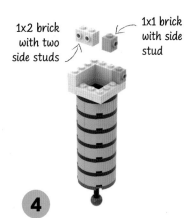

- 1x2 brick with two side studs
- 1x1 brick with side stud

4

Build a square shape onto the end using plates and bricks with side studs. Leave a gap two studs wide on one side.

- 1x1 brick with studs on two sides

5

Add two plates and a row of bricks to two sides. Include bricks with studs on two sides in the corners.

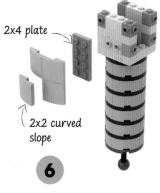

- 2x4 plate
- 2x2 curved slope

6

Connect a 2x4 plate to the side studs. Add four 2x2 curved slopes on top of it.

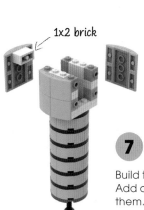

- 1x2 brick

7

Build two more rounded sides. Add a 1x2 brick to one of them. Connect both rounded sides to the side studs.

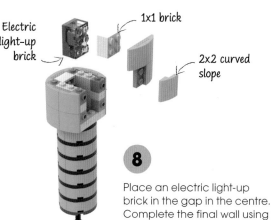

- Electric light-up brick
- 1x1 brick
- 2x2 curved slope

8

Place an electric light-up brick in the gap in the centre. Complete the final wall using a 2x2 brick, a plate, and slopes.

The clapper for this clapperboard frame moves on a hinge plate.

This fish piece connects to the frame via a plate with clip.

Life hack!

As well as photos, these frames can be used to display funny drawings or notes. They will stop the edges from getting ragged, too.

PHOTO FRAMES

★ ★ ★ ★ ★ ★ ★ ★ ★

My friend is already a star!

Get your head in the frame! Show off a favourite photo in a themed frame. A holiday snap could have a beach-themed frame. How about a clapperboard frame for a photo of a friend who wants to be a movie star? Just slide your photo into the slot and find the perfect spot to display it!

How to build:

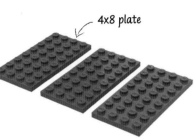

4x8 plate

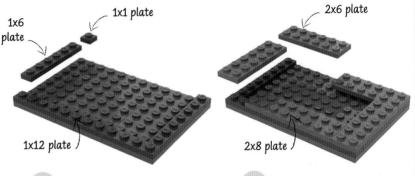

1x6 plate

1x1 plate

1x12 plate

2x6 plate

2x8 plate

1

Start with the back of your frame. Lay three 4x8 plates next to each other.

2

Now build the second layer. Place plates around the edge of your frame, leaving a gap at the top for your photo.

3

Next create a third layer. Make it one stud wider than the previous layer so the photo will stay in place.

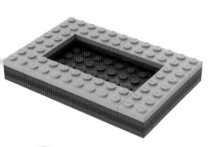

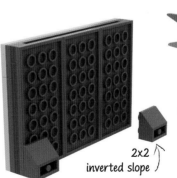

2x2 inverted slope

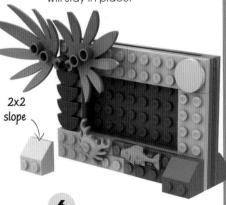

2x2 slope

4

Repeat with one more layer around the edge, to give your frame some depth.

5

Add two bricks with slopes to the back of your frame. These will provide support so that your frame doesn't tip over.

6

Decorate your frame. Make sure the decorative pieces don't cover your photo. Add two more slopes for support.

It's cool to be square!

MOSAICS

★ ★ ★ ★ ★ ★ ★ ★ ★ ★

No one wants to see scratches on their furniture, whether it's flat-pack or family heirloom. Protect yours with these stylish LEGO mosaics. Slip one under an ornament, a trinket dish, or your bedside alarm clock and never worry about scuffs or marks again.

How to build:

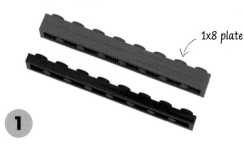

1x8 plate

1

Start with a 1x8 plate. This is the base of the frame. Then add two blue 1x8 plates on top to start off the design.

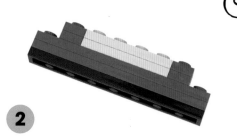

2

Place two yellow 1x4 plates in the centre, followed by two 1x1 plates on either side.

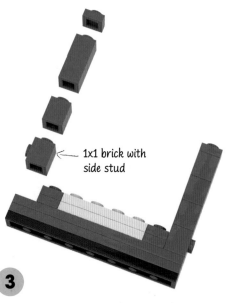

1x1 brick with side stud

3

Continue building up the minifigure face design. Include bricks with side studs facing outwards on either side to let you connect sideways tiles.

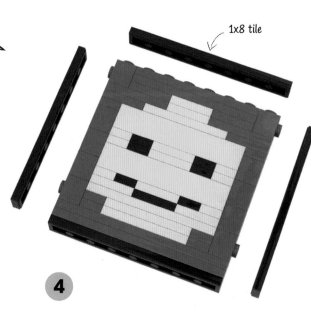

1x8 tile

4

To ensure your mosaic is exactly square, build the coloured layers to the height of six bricks and two plates. Add 1x8 tiles to the top and sides for smooth edges.

This design is made by rotating headlight bricks, allowing it to have tiles on all four sides.

This method involves building up in one direction from a single plate.

Use your mosaics to express yourself. If you love hot chocolate, why not create a mug design?

DAILY REMINDERS

★ ★ ★ ★ ★ ★ ★ ★ ★ ★ ★ ★ ★

Take yourself to task with a bedside memory-jogger. Each night, add small builds to remind you of the next day's jobs, such as washing your jeans or finishing your homework. In the morning, it's the first thing you'll set eyes on.

Life hack!

The stand for your reminders is elevated so that it won't take up too much room on your bedside table. Your phone, a book, or a few trinkets can fit underneath it.

This t-shirt build could remind you to iron your favourite top.

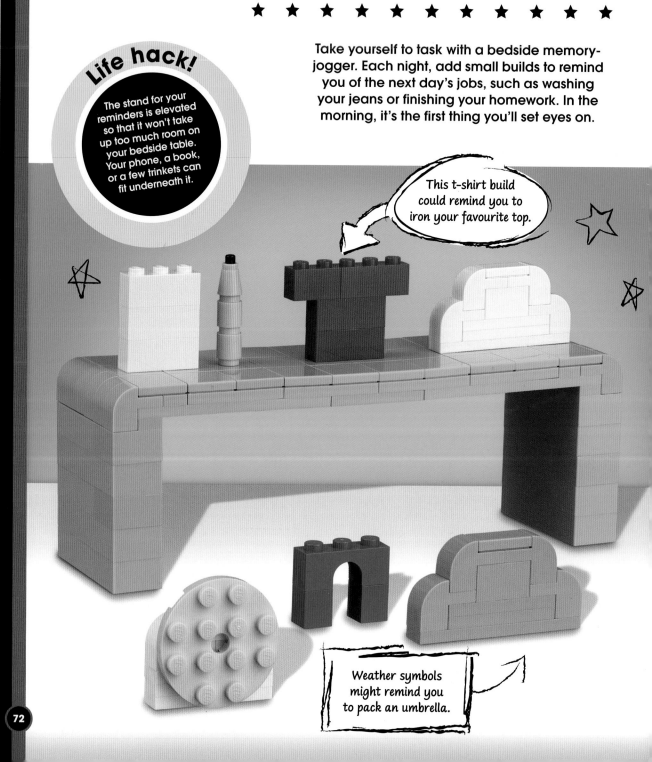

Weather symbols might remind you to pack an umbrella.

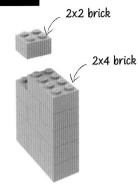

2x2 brick

2x4 brick

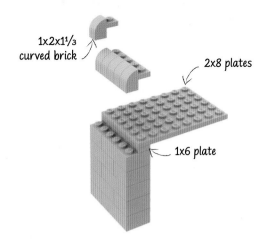

1x2x1⅓ curved brick

2x8 plates

1x6 plate

1 Stack two-stud-wide bricks until you have a wall five bricks high and six studs long.

2 Add a 1x6 plate and then three 2x8 plates on top. Place six 1x2x1⅓ curved bricks along the outside edge.

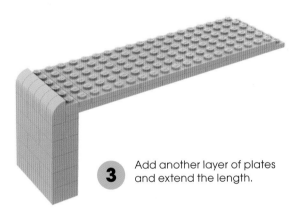

3 Add another layer of plates and extend the length.

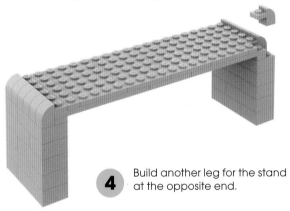

4 Build another leg for the stand at the opposite end.

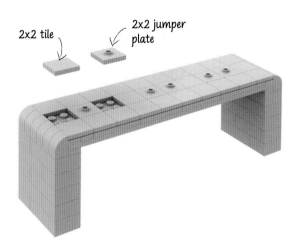

2x2 tile

2x2 jumper plate

5 Cover the plates with 2x2 tiles and 2x2 jumper plates. The jumper plates will hold your reminder builds in place.

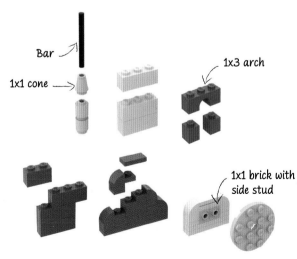

Bar

1x1 cone

1x3 arch

1x1 brick with side stud

6 Build your reminder icons. Keep them simple and make sure they are the right size to fit onto the jumper plates.

MINIFIGURE DISPLAY FRAMES

★ ★ ★ ★ ★ ★ ★ ★

Put a minifigure squarely in the limelight with its own display frame. A sports star can stand proudly on a simple podium. An explorer will look at home framed by vines and spiders. Why not add a background too, like this mermaid's bubbly blue one?

How to build:

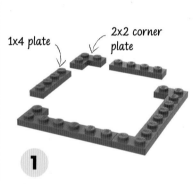

1x4 plate
2x2 corner plate

1
Create a square frame using plates, with corner plates at each corner.

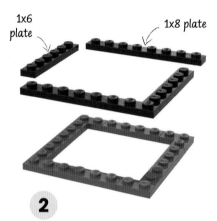

1x6 plate
1x8 plate

2
Build another layer of plates over the top so that the first layer of plates is secured.

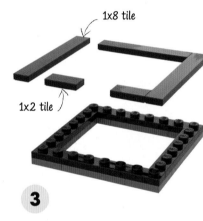

1x8 tile
1x2 tile

3
Add another layer of plates or tiles. Leave a two-stud-wide gap on one side.

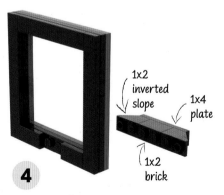

1x2 inverted slope
1x4 plate
1x2 brick

4
Make a stand made from a brick, two slopes, and a plate. Connect it to the back of the frame.

1x2/1x2 angle plate

5
Add an angle plate to the gap in your frame. This is where your minifigure will stand.

This build is a winner!

Creepy flourishes top off this explorer's frame.

Secure larger accessories to the frame with clips.

I'm in my element!

Create a larger surface using plates and wedges so there is more room to display accessories.

LOCKABLE SAFE

★ ★ ★ ★ ★

Have you ever lost your pocket money and wondered "Hmm... have I really lost it, or has a sneaky someone "borrowed" it?" Remove those moments of doubt by storing money and other valuables in this LEGO safe.

Squawk! Keep your beaks out!

The key's ornate golden handle is a LEGO® NINJAGO® piece.

Inspirational build!

Perfect for hiding your treasure!

This plate with ring is integral to this build. When the key is turned, it pushes the long axle below into the ring, locking the lid in place.

Life hack!
Keep the key on you at all times by threading the handle over your favourite necklace or hanging it from your belt loop.

LEGO Technic axle

Turning the key while the lid is down locks the safe.

CHAPTER 4
★ ★ ★ ★ ★ ★ ★ ★
STATIONERY HACKS

The perfect spot for my paintbrush!

Utensil holder

Tape dispenser

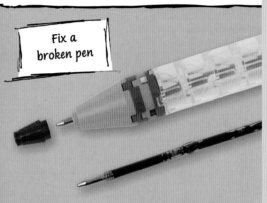

Fix a broken pen

Paperweight

Paperclip holder

STATIONERY POT

★ ★ ★ ★

Any pot can hold pens and pencils. This LEGO® pot offers something extra – it can hold smaller items of stationery, too. Little drawers at the base pull out to reveal pencil sharpeners, paper clips, and erasers.

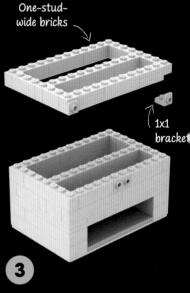

One-stud-wide bricks

1x1 bracket

How to build:

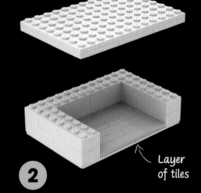

Layer of tiles

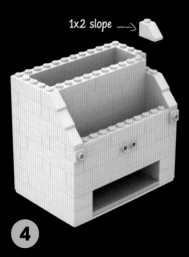

8x12 plate

1 Start with an 8x12 plate and then add walls to the back and sides using two-stud-wide bricks.

2 Add a layer of tiles to the plate. Add two layers of plates over the top of the bricks.

3 Build up the edges of the pot, with a row of bricks in the middle. Include a brick with side studs and 1x1 brackets.

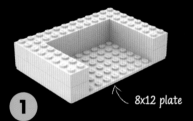

1x2 slope

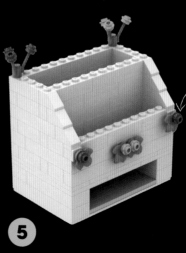

Leaf and flower

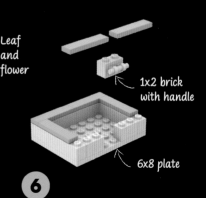

1x2 brick with handle

6x8 plate

4 Add another layer of bricks to the back, middle, and sides. Place a 1x2 slope at each end. Continue for three more layers.

5 Connect leaf and flower pieces to the side studs and brackets you built in earlier. Add grass and flowers.

6 Create a drawer out of a 6x8 plate and one-stud-wide bricks and tiles. Include a brick with a handle at the front.

Why not build some fun LEGO stationery to decorate your pot?

Life hack!

Dividing your pots into sections makes it easier to keep your stationery organized. Large items like scissors can go in the tallest section, and stumpier pencils sit at the front.

The layer of tiles below the drawers allows them to slide in and out smoothly.

This build is a masterpiece!

PAPERCLIP HOLDER

★ ★ ★ ★ ★ ★ ★ ★ ★

These are the bee's knees...

Don't keep your paper clips in a pot. If it gets knocked over you'll spend ages picking them up off the floor. Try one of these fun magnetic LEGO holders instead, and make sure those pesky paper clips stick around.

How to build:

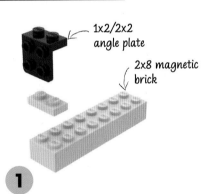

1x2/2x2 angle plate

2x8 magnetic brick

1

Take a 2x8 magnetic brick and add a 1x2 plate to one end, then an angle plate over the plate.

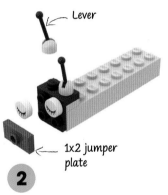

Lever

1x2 jumper plate

2

Use two lever pieces to make bee antennae. Next add eye pieces and a pink 1x2 jumper plate for a face.

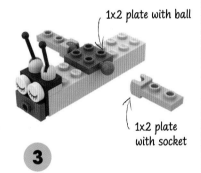

1x2 plate with ball

1x2 plate with socket

3

Place two 1x2 plates with a ball in the centre of the magnetic brick. Connect a 1x2 plate with socket to each.

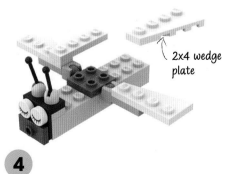

2x4 wedge plate

4

For wings, add two white 2x4 wedge plates on either side. Place two more wedge plates on top, at right angles.

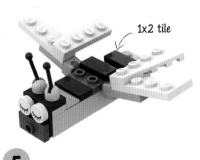

1x2 tile

5

Finally, add black stripes to your bee using 1x2 tiles spaced out evenly. Place the wings at an angle.

PAPERCLIP

★ ★ ★ ★ ★ ★ ★ ★ ★ ★ ★

Get a grip! This LEGO paperclip will keep notes, tickets, stamps, and passes firmly together for when you need them. So gather up those stray pieces of paper – this clip's jaws are poised for the pinch!

> This LEGO elastic band piece fits over the plates and is tight enough to hold them firmly together.

How to build:

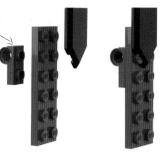

1x2 plate with pin hole underneath

1

Add a tile with pointed edge and a plate with a pin hole to an 8x2 plate. Repeat so that you have two identical structures.

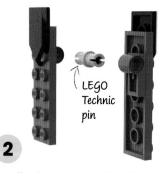

LEGO Technic pin

2

Turn the two sides back to back. Secure them together by adding a LEGO® Technic pin to the two pin hole pieces.

LEGO elastic band

3

Give your paperclip grip power by placing a LEGO elastic band around the thin end.

BROKEN PEN

★ ★ ★ ★

We all know the dreaded sound of a biro breaking. If it's only the plastic that's damaged, don't throw the pen away. Instead, try this ink-redible LEGO hack. Your biro will work as well as before – and it will look much smarter!

Life hack!

Any colour of bricks will work for this build, but if you use transparent bricks you'll be able to see when the ink is running low!

Make sure the ink tube is no wider than 3 mm so that it can fit through the gap.

How to build:

2x2 round brick

1

Stack seven 2x2 round bricks together to form a cylinder shape. Slide the ink tube through the centre.

1x1 cone

2x2x2 cone

2

Add three 2x2 round plates over the ink tube. Add a 2x2x2 cone piece over the end, then a 1x1 cone to act as a pen lid.

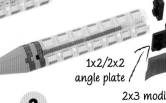

1x2/2x2 angle plate

2x3 modified brick

3

Secure the other end of the pen with plates. Make a clip with an angle plate and a 2x3 modified brick.

The roll of tape sits loosely so that it can rotate.

The end of the tape can be stuck here. Tear off pieces against the edge of the bricks.

Decorate your holder with colourful pieces. This example is decorated to look like a brightly wrapped gift.

Life hack!

This model works best with masking tape, which can be torn easily. Use the edge of the build as you would use the sharp edge of a regular tape dispenser, to tear off pieces of tape.

TAPE DISPENSER

★ ★ ★ ★ ★ ★ ★ ★ ★ ★ ★ ★

Are you ready to roll? Then you'll love this clever LEGO tape dispenser. It lets you unwind just the right amount of sticky tape for your crafting needs. When you've torn off the tape, stick the free end to the build, ready for next time.

Perfect for present wrapping!

How to build:

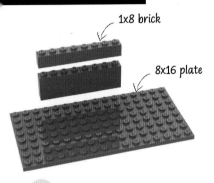

1x8 brick

8x16 plate

1

Start with an 8x16 plate and add bricks to the back.

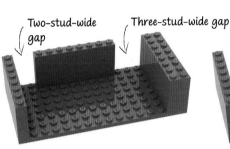

Two-stud-wide gap

Three-stud-wide gap

2

Add more bricks to the sides.

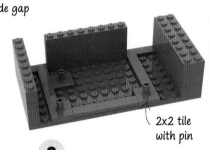

2x2 tile with pin

3

Add tiles and three 2x2 tiles with pins as shown above.

18 mm diameter wheel

11 mm diameter wheel

4

Place an 18 mm diameter wheel on one of the pins, and 11 mm diameter wheels on the other two pins.

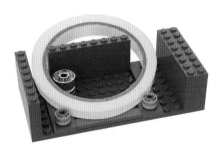

5

Place the tape into the holder so it rests between the wheels. Stick the end of the tape to the wide edge of the holder.

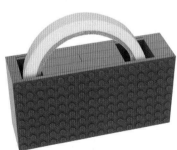

6

Complete the tape holder by adding another 8x16 plate to the open side.

87

UTENSIL HOLDERS

★ ★ ★ ★ ★ ★ ★ ★ ★ ★ ★

I never forget a handy hack!

When working on an art project, have a few helpers standing by to hold your tools. This giraffe will protect the point of a sharpened pencil, and the elephant will hold a brush safely out of the way.

How to build:

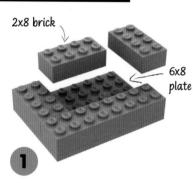

2x8 brick

6x8 plate

1

Create the base for your build by placing bricks around the edge of a plate. Leave a gap in the middle.

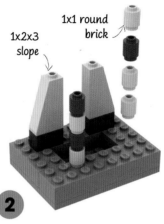

1x1 round brick

1x2x3 slope

2

Start building the giraffe's legs. Use slopes on top of bricks for the hind legs, and stack 1x1 round bricks for the front legs.

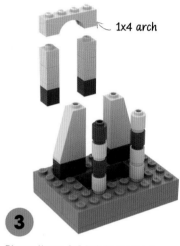

1x4 arch

3

Place three 1x1 bricks behind each back leg. Add a 1x4 arch over the top. This will be the base for the giraffe's neck.

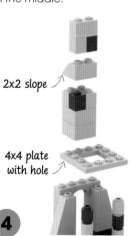

2x2 slope

4x4 plate with hole

4

Connect a 4x4 plate with hole to the four legs. Build the neck up, placing a 2x2 slope in the middle.

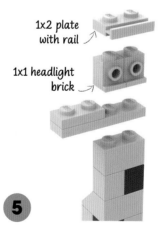

1x2 plate with rail

1x1 headlight brick

5

Add a 1x4 plate, then a 1x2 plate with two 1x1 headlight bricks next to it. Top these with a 1x2 plate with rail.

Tap

1x1 round plate

6

Finish with black tap pieces for horns and two white 1x1 round plates for teeth. Connect eye pieces to the headlight bricks.

The elephant's tusks are perfect for cradling a paintbrush. What other pieces might perform the same function?

Life hack!

Your holders don't have to be animal themed. Come up with a theme that suits you, or simply build them in your favourite colours.

As well as art supplies, your builds can hold chopsticks, pens, a nail file, or makeup brushes.

PAPERWEIGHTS

★ ★ ★ ★ ★ ★ ★ ★ ★ ★ ★ ★ ★ ★ ★ ★

"Weight" for me!

Fans, open windows, or the draught from someone rushing past can send notes and papers fluttering all over the place. Holding them down is a breeze with these LEGO paperweights. Build it, and save yourself any more pointless paperchases.

How to build:

2x8 plate

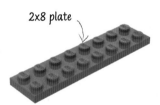

1

Build your vehicle from the bottom up. A piece like this 2x8 plate forms a sturdy base for the racing car.

1x4 brick with side studs

1x4 slope

2

Add 1x4 bricks with side studs to each side of the plate and 1x4 slopes for a bonnet. Use any colours you like.

2x3 plate with hole

2x2 slope

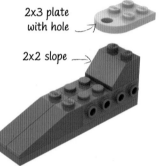

3

Streamline your car with a spoiler. Stack a 2x2 slope and a 2x3 plate with hole at the rear of your racing car.

1x2 plate

1x1 slope

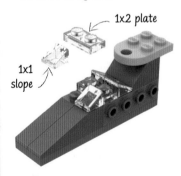

4

Use transparent pieces such as 1x1 slopes and 1x2 plates to make shiny windows for your vehicle.

Radar dish

1x4 plate with two studs

5

Fix radar dishes to a plate with two studs. Attach the plate to the brick with side studs. Be sure to add wheels on both sides.

Life hack!

Build your paperweight in whatever shape you like. If you have a lot of papers to weigh down, you may find a chunkier design, such as an avocado half, works better.

Dark green tiles, light green plates, and round brown pieces make a vibrant avocado.

The slim racing car won't take up too much space on a cluttered desk.

NOTE HOLDER

★ ★ ★ ★ ★ ★ ★ ★ ★

Jotting things down is a great habit. It helps you to remember important little tasks and records all your clever ideas, too. Keep a stack of sticky notes close by, and let this eye-catching, upstanding penguin hold it for you.

Try this cool idea!

Life hack!

Have fun creating your own designs – just make sure there are protruding pieces to hold the notepad in place. Space them out at the correct height so your notepad fits snugly.

The notepad is held between the penguin's beak and feet.

Put the holder somewhere you will see it, like on your desk or bedside table.

How to build:

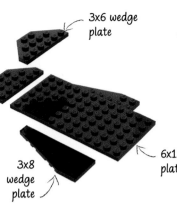

3x6 wedge plate

3x8 wedge plate

6x12 plate

1

Create the shape of the penguin's body using a rectangular plate and four wedge plates.

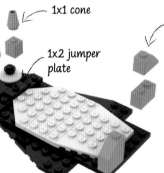

1x4 wedge plate

2

Add white plates to the middle. Use plates with corner cutouts to give a more rounded shape.

1x1 cone

1x2 slope

1x2 jumper plate

3

Build eyes with 2x2 round dishes and 1x1 round plates, and a beak with a cone and brick. Connect these to the penguin using jumper plates. Add feet built from bricks and slopes.

4

Finally, add a stand to the back of your note holder to give it stability.

RULER
★ ★ ★ ★ ★ ★

This snake ruler nestles in a drawer, ready to uncoil whenever you need to draw a straight line or measure your minifigure! Thirty centimetres is about the right scale for your anaconda ally.

Honestly, the lengths I go to!

Hinge plates allow the snake to curl up and protect its tail.

Life hack!

Space out the hinge pieces so that the snake can fold neatly. Try laying out the bricks in the curled up position before you secure them and add the hinges.

93

Fidget cube

Laundry lowerer

Bin hoop

Craft some crazy contraptions!

Kaleidoscope

Press here to launch your paper into the bin!

Ready... aim...fire!

The inverted radar dish holds a crumpled-up piece of paper, ready to fly!

CATAPULT

★ ★ ★ ★ ★ ★ ★ ★ ★ ★

*Once more
unto the bin!*

Turn disposing of wastepaper into a game of skill
with this clever catapult. Hit the button to flip the
lever and launch your screwed-up scraps through
the air and into the waiting bin. Imagining your bin
as an enemy fortress only adds to the fun!

How to build:

Axle pin

1x2 brick
with hole

Handle

1

Build a wall with a sloped front
onto a 16x16 baseplate. Add an
axle pin and handle attached
to a 1x2 brick with hole.

Leave a gap of
1x2 on either side

2

Add another wall so it sits four
studs apart from the first one on
the baseplate. Leave a 1x2 gap
in each wall, facing inwards.

4x4 round
plate and
2x2 tile

8x8 radar
dish

Click hinge

3

Build a long lever using plates,
with an 8x8 radar dish at one
end. Add two bricks with holes
to connect to the walls.

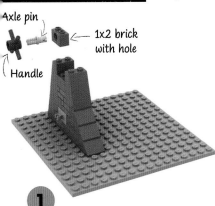

Axle

1x2 brick
with axle hole

Axle connector

4

Add an axle and axle
connectors to either side via
the bricks with holes. Then
add bricks with axle holes.

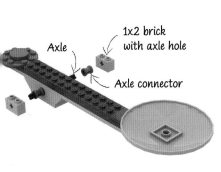

5

Connect the catapult to the
two walls by placing the side
bricks into the gaps you
created earlier.

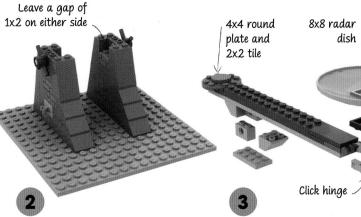

Fang

6

Add more bricks to the
sides. Include plates with clips
to attach decorative details
like these fang pieces.

POINTER

★ ★ ★ ★ ★ ★ ★ ★ ★

Let's be honest – getting out of bed to turn off the main light is a bore. If that light switch is just out of reach, a LEGO® pointer can help. Reach out with the pointer, click the switch, and let blissful darkness descend. Goodnight!

3, 2, 1... we have light off!

How to build:

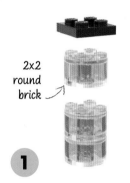

2x2 round brick

1

Stack three 2x2 round bricks. Then place a 2x2 plate on the top of your round brick stack.

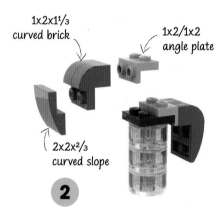

1x2x1⅓ curved brick

1x2/1x2 angle plate

2x2x⅔ curved slope

2

Attach angle plates to each side of the 2x2 plate. Curved bricks and slopes create the familiar rocket fin shape.

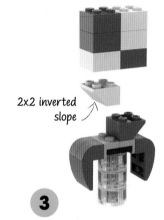

2x2 inverted slope

3

Add two 2x2 inverted slopes on top of the angle plates. Form the body of the rocket ship with a stack of 2x2 bricks.

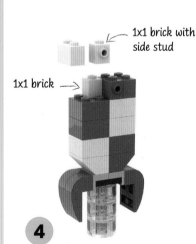

1x1 brick with side stud

1x1 brick

4

Place 1x2 bricks on each side of the rocket and 1x1 bricks at the back. Add bricks with side studs with studs facing forwards.

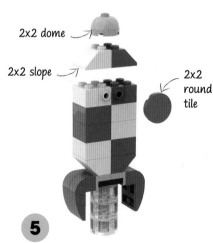

2x2 dome

2x2 slope

2x2 round tile

5

Create a streamlined shape with 2x2 slopes. Use a 2x2 dome to make a nose cone and a round tile for a window.

Axle

6

The rocket is ready! Slide an axle through the stack of 2x2 round bricks. The axle is your pointer's handle.

KALEIDOSCOPE

★ ★ ★ ★ ★ ★ ★ ★ ★ ★ ★ ★ ★ ★ ★ ★ ★ ★

Bricks ahoy!

With this kaleidoscope, patterns are created when loose LEGO pieces collide. Perhaps it should really be called a collide-o-scope! Look through one end of the kaleidoscope and rotate the other to see an ever-changing pattern of gem-like shapes.

Look through the hole at this end.

These two transparent 1x6x5 panels filled with colourful bricks rotate on a pin.

How to build:

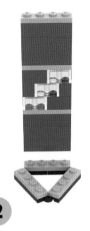

1x4 plate

1x4 hinge plate

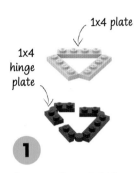

1

Arrange three 1x4 hinge plates in a triangle shape. Place three 1x4 plates over the top.

2

Build a one-stud - wide wall on one side of the triangle.

3

Build another wall and connect it to the adjacent side.

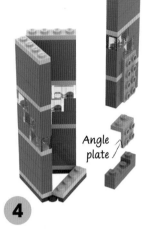

Angle plate

4

Add a third wall. Build in angle plates between the layers so there are studs on the side of the wall.

1x1 slope

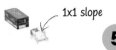

5

Lock the walls in place with three more hinge plates. Place transparent plates and slopes on top of each wall to create a smooth surface.

6

Cover the angle plates with bricks. Add two 1x10 curved slopes to the bricks and place a 2x2 tile over these.

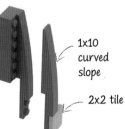

1x10 curved slope

2x2 tile

1x6x5 panel

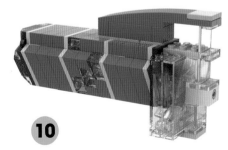

7

2x3 plate

Take two 1x6x5 panels. Place small colourful pieces in between these and secure with a 2x3 plate.

1x2x2 panel

1x2 brick with hole

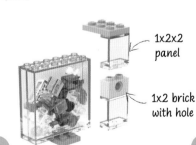

8

Place a 1x2x2 panel on the edge of the bottom plate. Add a brick with hole, then another panel. Fix with a 2x3 plate at the top.

Pin

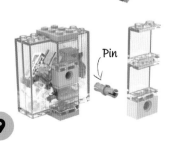

9

Connect another brick with hole using a pin. Stack two 1x2x2 panels on top. The pin will allow the panels to spin around.

10

Fix the transparent section to the main section by connecting the top 1x2x2 panel to the curved slopes.

CHAIN REACTION

★ ★ ★ ★ ★ ★ ★ ★ ★ ★ ★ ★

The brick-built domino knocks into the windmill's sail, making it spin.

The minifigure knocks over the hammer, which triggers the shooting function on the vehicle.

I'm giving it my best shot!

There are two ways of doing a job – the simple way and the fun, creative way. Challenge yourself to build a chain reaction machine to do a simple job. This series of small builds forms an amusingly complex system for replacing a pen lid.

Inspirational build!

The windmill's sail pushes the ball down this ramp, knocking over the minifigure at the other end.

I do all my own stunts, you know.

The missiles flick the pen lid, which then falls down onto the pen.

Life hack!

Experiment with builds that have simple functions, whether that's a hinge that opens and closes or a rolling ball that knocks something over. How can these functions combine to create a chain reaction?

BIN HOOP

★ ★ ★ ★ ★ ★

Slam dunk that paper junk with a LEGO basketball hoop. Fix the hoop and backboard to the edge of a wastepaper bin, then take aim with your screwed-up notes, doodles, and snack wrappers. Too easy? Try placing the bin further and further away!

How many baskets can you score?

How to build:

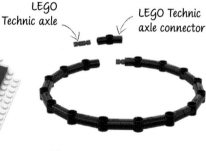

4x4 wedge plate

4x12 plate

1

Create a large rectangle shape using plates and wedge plates connected by narrower plates at the back.

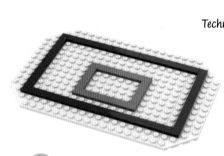

2

Add red and black tiles to form two concentric rectangles.

LEGO Technic axle

LEGO Technic axle connector

3

Build a circle using LEGO® Technic axle connectors joined by axles.

LEGO Technic pin

1x2 brick with holes

2x4 plate

1x4 brick

4

Attach two 1x2 bricks with holes to a 1x4 brick using a 2x4 plate. Lie it sideways so that the holes face up. Add pins to two holes and connect the circle.

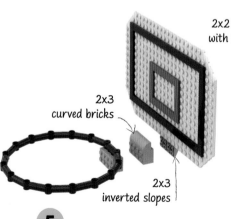

2x3 curved bricks

2x3 inverted slopes

5

Add inverted slopes to the back of the hoop. Connect the two main sections via curved bricks and a plate.

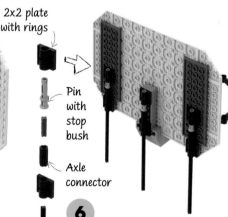

2x2 plate with rings

Pin with stop bush

Axle connector

6

Connect two plates with rings, a pin with stop bush, an axle, and an axle connector to a long axle. Add these to the back.

Aim the wastepaper at the red rectangle so it bounces into the bin.

Life hack!

If you want to make this more of a challenge, build the hoop with a smaller diameter so that you have to aim your paper more carefully.

The basketball hoop sits over your wastepaper basket ready for you to aim your paper at.

105

GRABBER

★ ★ ★ ★ ★ ★ ★ ★ ★ ★

If you've got chores, you need these jaws! This extendible grabber makes tidying your room fun – no more bending down to pick things up off the floor. Squeeze the handles and the shark's head shoots out to snap up papers, toys, and unlucky minifigure victims.

Here's something to get your teeth into!

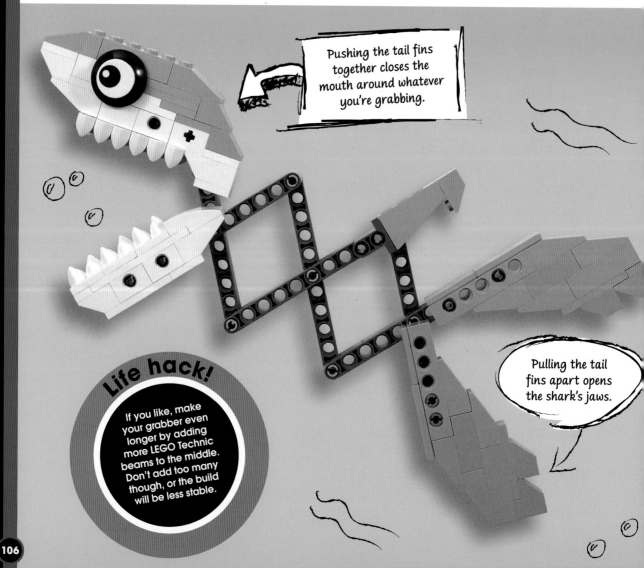

Pushing the tail fins together closes the mouth around whatever you're grabbing.

Pulling the tail fins apart opens the shark's jaws.

Life hack!

If you like, make your grabber even longer by adding more LEGO Technic beams to the middle. Don't add too many though, or the build will be less stable.

How to build:

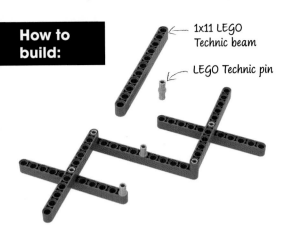

1x11 LEGO Technic beam

LEGO Technic pin

1 Lay six 1x11 LEGO® Technic beams in a criss-cross pattern. Secure them with LEGO Technic pins.

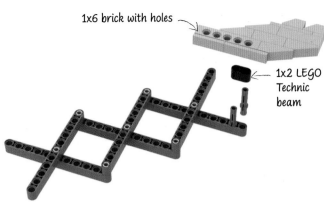

1x6 brick with holes

1x2 LEGO Technic beam

2 Build a tail fin and include a 1x6 brick with holes. Connect this using a 1x2 beam and long pins.

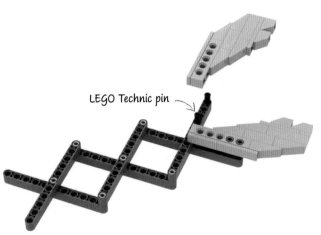

LEGO Technic pin

3 Build another tail fin, again with a 1x6 brick with holes. Attach to the beams using two pins.

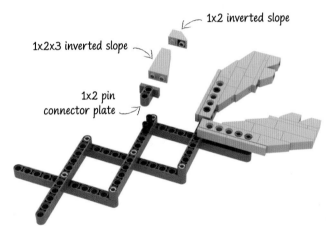

1x2 inverted slope

1x2x3 inverted slope

1x2 pin connector plate

4 For a fin, connect two inverted slopes using a 1x2 pin connector plate and two pins.

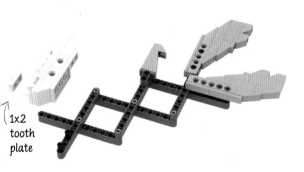

1x2 tooth plate

5 Build a lower jaw shape with rows of tooth plates. Build in two bricks with holes and attach these with pins.

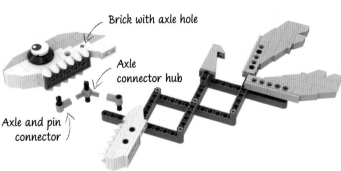

Brick with axle hole

Axle connector hub

Axle and pin connector

6 Build a head and attach it using an axle connector hub and two axle and pin connectors.

PADLOCK

★ ★ ★ ★ ★ ★ ★ ★ ★ ★

Lock your things away from prying eyes with this chunky LEGO padlock. A scary minifigure head on the key acts as a warning against tampering with the lock. Of course, any snooper could just break the bricks apart but shhh... don't tell them that!

Now no-body can steal your stuff!

Life hack!

This padlock could be used in a variety of ways. You could place it around your wardrobe door handles or use it to secure the pages of your diary.

A LEGO Technic axle connects to a brick with holes.

The padlock opens with the help of a "skeleton key" topped with a minifigure skeleton head!

How to build:

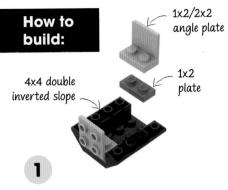

1x2/2x2 angle plate

1x2 plate

4x4 double inverted slope

1

Start with a 4x4 double inverted slope and add a 1x2 plate and angle plate at either end.

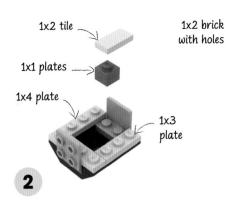

1x2 tile

1x1 plates

1x4 plate

1x3 plate

2

Place plates around the edge of the inverted slope, leaving a gap at one end. Add 1x1 plates and a 1x2 tile.

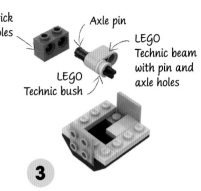

1x2 brick with holes

Axle pin

LEGO Technic beam with pin and axle holes

LEGO Technic bush

3

Place a LEGO Technic bush and beam with pin and axle holes onto an axle pin. Put the pin into a brick with holes.

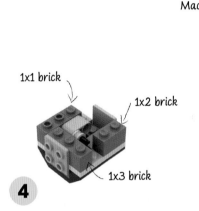

1x1 brick

1x2 brick

1x3 brick

4

Attach the brick with holes, then a 1x1 brick beside it. Add a 1x3 brick and a 1x2 brick, leaving a gap opposite the axle pin.

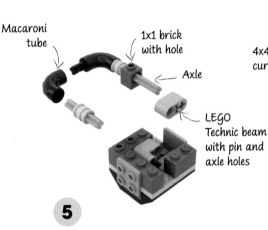

Macaroni tube

1x1 brick with hole

Axle

LEGO Technic beam with pin and axle holes

5

Connect two macaroni tubes. Add axles, bushes, a brick with hole, and a Technic beam with pin and axle holes.

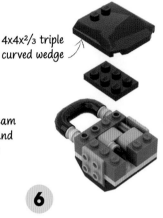

4x4x²/₃ triple curved wedge

6

Connect the section from the previous step to the brick with holes. Finish the padlock with a plate and a curved wedge.

How it works:

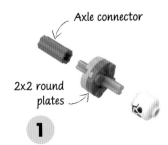

Axle connector

2x2 round plates

1

Build a key by placing a minifigure head, two round plates, and an axle connector onto an axle.

2

The end of the axle connector attaches to the end of the axle pin inside the padlock.

3

Turning the key rotates the beam, which is holding the padlock shut. This allows the top of the padlock to lift up.

ZIP WIRE

★ ★ ★ ★ ★

"Hey, pass me that pen", someone says. Instead of moaning "Get up and get it yourself", you smile smugly, slip the pen into the pod, and send it whisking off to them on this LEGO zip wire. And neither of you will have to move from the comfort of your chairs!

Get your goodies from A to B!

How to build:

1

Build around the edges of an 8x8 plate with three layers of one-stud-wide bricks.

1x2 plate with pin connector

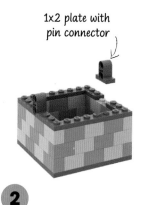

2

Add plates along the edges, leaving space for two 1x2 plates with pin connectors at opposite ends.

1x11 LEGO Technic beam

LEGO Technic pin

2x2 plate with ring underneath

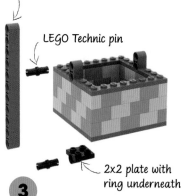

3

Add a 1x11 LEGO Technic beam using a pin. Connect the beam at the bottom using a pin and a plate with ring.

2x2 round tile with hole

2x2 round corner tile

4x4 round plate

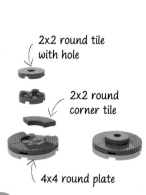

4

Add 2x2 round corner tiles, a 2x2 round plate, and a 2x2 round tile with hole to two 4x4 round plates.

Axle

LEGO Technic bush

5

Connect these to the beam using a long axle and a LEGO Technic bush. Place another bush on the other end.

6

Connect another beam to the opposite side using two pins and a 2x2 plate with ring underneath.

FIDGET CUBE

★ ★ ★ ★ ★

Do you tend to fiddle with a pen or bite your nails when you're bored, anxious, or thinking hard? This LEGO fidget cube is a much better stress reliever. Its levers, studs, and wheels will keep your hands busy – and busy hands mean a calm mind.

Time to shake, rattle, and roll!

Rotate this plate with your thumb.

Wobble the spring piece back and forth.

Slide this section up and down.

Spin these two tyre pieces.

Press this ball piece up and down.

Flick this hinge back and forth.

How to build:

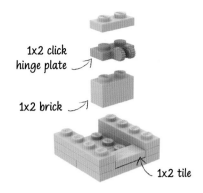

1x2 click
hinge plate

1x2 brick

1x2 tile

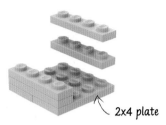

2x4 plate

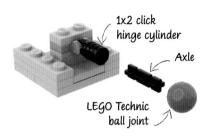

1x2 click
hinge cylinder

Axle

LEGO Technic
ball joint

1 Connect two 2x4 plates with four 1x4 plates.

2 Add a 1x2 tile to the front. Stack a 1x2 brick, a click hinge plate, and a 1x2 plate and add to the centre.

3 Connect a click hinge cylinder, an axle, and a ball joint to the click hinge plate.

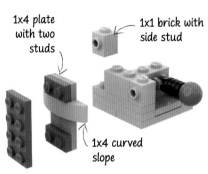

1x4 plate with two studs

1x1 brick with side stud

1x4 curved slope

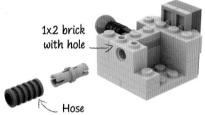

1x2 brick with hole

Hose

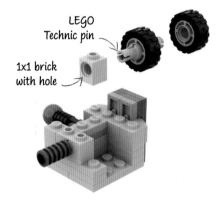

LEGO Technic pin

1x1 brick with hole

4 Line up two 1x1 bricks with side studs and a 1x2 brick. Add plates and a curved slope to the side studs.

5 Turn the build around. Add a 1x1 brick and a 1x2 brick with hole to this side. Attach a hose to the hole using a LEGO Technic pin.

6 Connect two tyre pieces to a 1x1 brick with hole using a pin. Place the brick with hole in the corner.

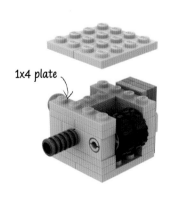

1x4 plate

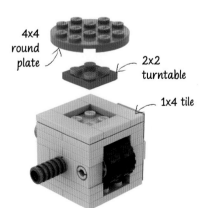

4x4 round plate

2x2 turntable

1x4 tile

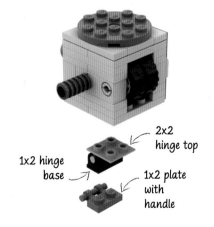

2x2 hinge top

1x2 hinge base

1x2 plate with handle

7 Build up the sides by two plates and add another two plates to the centre. Place two 2x4 plates on top.

8 Place tiles around the edges and then add a 2x2 turntable in the centre. Connect a 4x4 round plate.

9 Connect a hinge base to a 2x2 hinge top. Place on top of a 1x2 plate with a handle. Add these to the bottom.

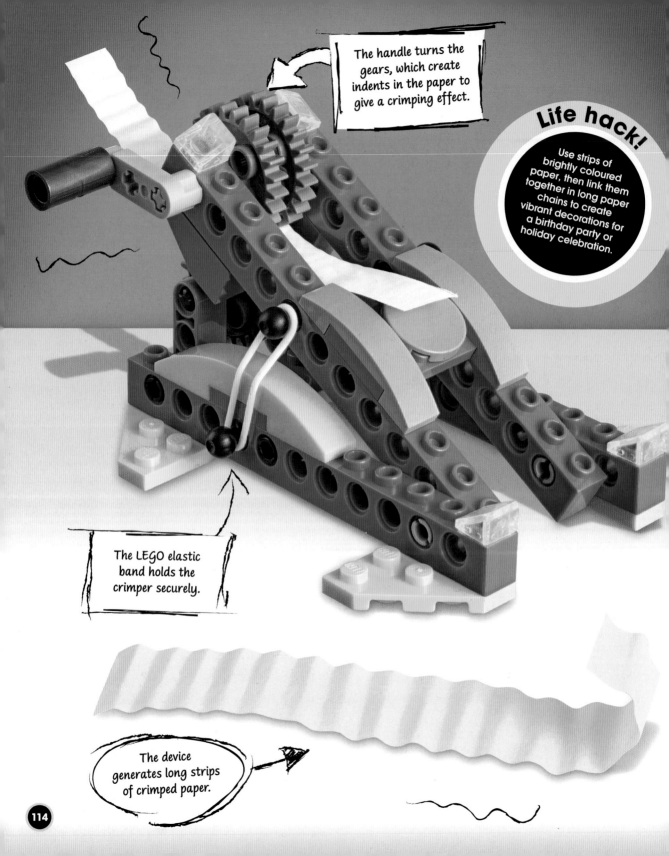

The handle turns the gears, which create indents in the paper to give a crimping effect.

Life hack!

Use strips of brightly coloured paper, then link them together in long paper chains to create vibrant decorations for a birthday party or holiday celebration.

The LEGO elastic band holds the crimper securely.

The device generates long strips of crimped paper.

PAPER CRIMPER

★ ★ ★ ★ ★ ★ ★

If you're a keen crafter, you'll know that little details
such as crimped paper can make a big difference.
What if you don't have crimped paper? Build a LEGO
contraption that will allow you to create some!

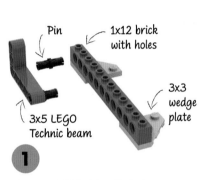

Pin
1x12 brick
with holes
3x3
wedge
plate
3x5 LEGO
Technic beam

1

Place a 1x12 brick with holes
onto two 3x3 wedge plates.
Connect a 3x5 LEGO Technic
beam using two pins.

1x3
inverted
slope

2

Add another 1x12 brick with
holes and a 1x3 inverted slope.
Place this section diagonally
and connect with a pin.

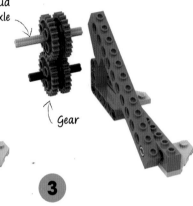

Five-stud
long axle
Gear

3

Place two gears over an axle,
and two more gears over a
shorter axle. Connect to the
holes in the sides.

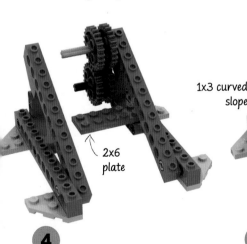

2x6
plate

4

Repeat steps 1–2, but with the
pieces mirrored, to create
the opposite side. Add a 2x6
plate between the two sides.

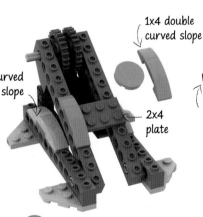

1x3 curved
slope
1x4 double
curved slope
2x4
plate

5

Add a 2x4 plate to the centre,
then two double curved slopes
and a round tile. Place two
1x3 curved slopes over each
end of the central 2x6 plate.

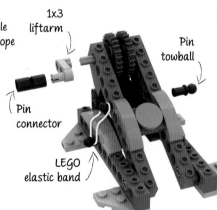

1x3
liftarm
Pin
towball
Pin
connector
LEGO
elastic band

6

Connect a 1x3 liftarm to the
top axle. Add a pin connector
to form a handle. Add two pin
towballs and an elastic band
on either side of the crimper.

115

LAUNDRY LOWERER

★ ★ ★ ★ ★ ★ ★ ★ ★

You're relaxing in your room when you hear a shout from the kitchen: "Any washing to go in?" Well, there's no need to bend down to pick up your dirty socks or walk down any stairs with them. Just use this labour-saving LEGO lowering device to pick up and shift those socks.

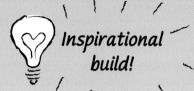

Inspirational build!

LEGO string piece wraps around winch.

Turn this handle to lower the claw, and then turn it the other way to draw it up again.

Life hack!

Securely attach a small basket to the end of this device using string. Now you can lower all kinds of items! Just tell people it's not laziness – it's efficiency.

MEET THE BUILDERS

★ ★ ★ ★ ★ ★ ★ ★ ★ ★

The models in this book were created by two talented LEGO® builders, Nate and Barney. Both are crazy about LEGO building! We asked them what it was like to build the models, and to share a few top tips...

Nate Dias

How many bricks do you own?
Hundreds of thousands, I couldn't possibly count!

What was your favourite model to build for this book?
I really like the classic space-themed bookends.

Do you know any good life hacks?
Pair your socks before putting them in the washing machine, so you never end up with missing socks.

Do you have any hacks that help with LEGO building?
I always have a notepad on me to jot down ideas for my next LEGO project.

Is there a particular task in your life you wish you could build a life hack for to make the job easier?
I wish there was a hack to make marking my students' books easier.

Which is your favourite brick?
My favourite LEGO element is a 1x2 plate. It's a very common piece, but it's amazing because it's the smallest piece that you can build really big things with.

"I always keep a long stick handy with some putty on the end..."

Barney Main

What was your favourite model to build for this book?
I love the phone speaker. It's just a box with a hole in the side but it works great; and the decorations are really fun!

Which model was the most challenging?
The chain reaction machine was tricky. Each section triggers the next one in the chain, so they all need to work together perfectly. It took a lot of trial and error – and endlessly setting up the dominoes – to get it right!

Do you know any good life hacks?
Don't have a colander? Use a cheese grater to drain peas and sweetcorn!

Do you have any hacks that help with LEGO building?
When building very large models, sometimes pieces fall down inside and you can't get them out. So I always keep a long stick handy with some putty on the end to help me pick them up!

Which brick did you find particularly useful for this book?
The 1x2 plate with ball is always useful! It's a stiff ball joint, so is great for holding stuff steady in any position. I also use it to anchor elastic bands.

BRICK GALLERY

When planning your builds, it can be useful to know which pieces you have and what they are called. Some pieces are particularly useful when it comes to creating functional models. Don't worry if you don't have all these parts – you can be creative with the pieces you do have.

★ ★ ★ ★ ★ ★ ★ ★ ★

Measurements

The width and length of pieces refers to the studs. Height is measured in bricks. Plates are shorter than bricks. Three plates are the same height as one brick.

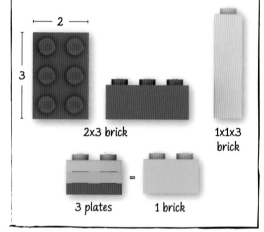

2x3 brick

1x1x3 brick

3 plates = 1 brick

Bricks

Bricks are the basis of most LEGO® builds. They come in many shapes and sizes, and are named according to size.

 Small parts and small balls can cause choking if swallowed. Not for children under 3 years.

2x2 brick

2x2 round brick

1x1 round brick

2x2 corner brick

2x2 round corner brick

2x2 dome

1x2 textured brick

1x1 cone

1x2x1⅓ brick with curved top

1x3 arch

1x3x2 inverted arch

Plates

Like bricks, plates come in many shapes and sizes. They are shorter than bricks – three stacked plates are the same height as one brick.

1x2 plate

1x2 jumper plate

1x4 plate with two studs

2x2 round plate

1x1 plate with ring

2x3 wedge plate

2x3 curved plate with hole

3x3 round corner plate

2x2 plate with pin holes

4x4 round plate with hole

1x1 plate with horizontal tooth

2x2 slide plate

Tiles

Tiles are the same height as plates, but are smooth on top instead of having studs.

2x2 tile   **2x2 round tile**   **1x1 quarter tile**

2x2 corner tile   **2x2 round corner tile**   **2x2 round tile with hole**   **2x2 printed tile**   **2x2 tile with pin**

Sideways building

Some bricks have studs on one or more sides, allowing you to build in multiple directions.

1x1 brick with side stud   **1x1 brick with two side studs**   **1x1 brick with four side studs**

Headlight brick   **1x2/2x2 angle plate**

Slopes

Slopes are bricks with diagonal angles. They can be big, small, curved, or inverted (upside-down).

1x2 slope   **1x2 inverted slope**   **1x3 curved slope**

Bars and clips

Any piece with a bar can fit onto a piece with a clip. Use clips and bars to make moving parts, such as crane arms.

1x1 plate with vertical clip   **1x1 plate with horizontal clip**

1x1 plate with top clip   **Bar holder with clip**

1x2 plate with bar   **1x2 plate with handle**   **Bar**

Moving pieces

Hinge plates allow builds to move side to side. Plates with sockets and plates with balls connect to form flexible joints.

 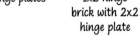

Hinge plates   **1x2 hinge brick with 2x2 hinge plate**

1x2 plate with ball   **1x2 plate with socket**   **1x2 click hinge**   **1x2 click hinge**

LEGO® Technic

LEGO® Technic parts expand the range of functions you can build into your models.

LEGO Technic pin   **LEGO Technic axle**   **LEGO Technic axle pin**   **LEGO Technic beam**   **LEGO Technic beam with pin and axle holes**

1x1 brick with hole   **1x2 brick with axle hole**   **1x2 pin connector plate**   **Axle connector**   **Axle and pin connector**   **LEGO Technic axle towball**   **LEGO Technic bush**

BUILD BASICS

★ ★ ★ ★ ★ ★ ★

The best thing about building with LEGO® bricks is using your creativity to build in any way you want. However, you might find it useful to follow a few simple guidelines. Read on to learn nifty techniques for creating incredible models.

Let's get building!

Keep it steady

If you want your model to stand up, it will need a stable base. Most freestanding builds start with a large plate to stop them from tipping over, while others stand on several bricks spread out like feet. As you build upwards, try not to put all the pieces on one side of the model. Spread the weight evenly and it won't wobble or fall.

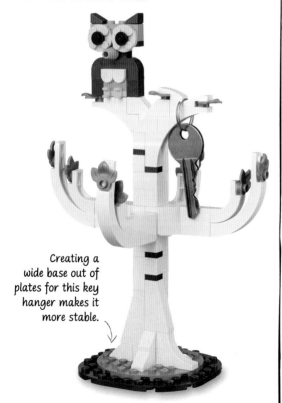

Creating a wide base out of plates for this key hanger makes it more stable.

These banana pieces are connected to bricks with hollow studs to make monster hair.

Make connections

Every single LEGO element can fit onto another in at least one way. Play around with a few unusual parts and you might be surprised by the ways you are able to connect them! You may have always used a piece in a certain way in the past, but that doesn't stop you from doing something completely different with it today!

Hold it together

A strong LEGO build contains lots of overlapping pieces at different sizes to hold everything together. Long bricks and plates are useful for locking smaller ones in place, while smaller pieces can lock themselves together if they are staggered (so that their sides do not line up). Corner bricks and plates make a strong connection where walls meet up, but again, only if they are overlapped with other parts.

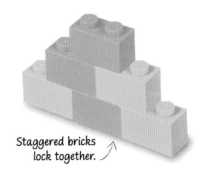

Staggered bricks lock together.

A 2x2 corner brick can overlap two bricks.

Think big – or small!

LEGO models can be built at any scale, from micro to mega! Micro-scale makes a few bricks go a long way, and uses small, unusual pieces in whole new ways. Building at life-size – or larger – makes a great challenge, and can lead to things you can really use!

The height of this model means your hat doesn't touch the ground.

This headphone wrap on page 40 is made of just nine pieces.

Think sideways

Whatever you want to build, think about whether it might look better if some or all of the parts were used on their sides – or even upside down! Angle plates and bricks with side studs offer ways to build in more than one direction. Even without these, you could make an entire model using only bricks laid down on their sides.

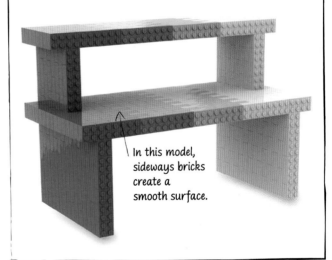

In this model, sideways bricks create a smooth surface.

Think ahead

If you want to tackle a bigger or more complex build, it can help to plan it out in advance or even draw some rough sketches. Think about the best bricks – and building methods – to achieve the effects you are looking for.

Cheers everyone!

Penguin
Random
House

Editor Rosie Peet
Project Art Editor Jenny Edwards
Production Editor Siu Yin Chan
Senior Production Controller Lloyd Robertson
Managing Editor Paula Regan
Managing Art Editor Jo Connor
Publisher Julie Ferris
Art Director Lisa Lanzarini
Publishing Director Mark Searle

Models created by Barney Main and Nate Dias
Photography by Gary Ombler

DK would like to thank: Randi K. Sørensen, Heidi K. Jensen, Robin James Pearson, Paul Hansford, Martin Leighton Lindhardt, Nina Koopmann, Charlotte Neidhardt, Henk van der Does and Peter Bernhard at the LEGO Group; Julia March for proofreading; and Tori Kosara, Helen Murray and Nicole Reynolds for editorial support.

First published in Great Britain in 2021
by Dorling Kindersley Limited
One Embassy Gardens, 8 Viaduct Gardens, London SW11 7BW

A CIP catalogue record for this book is available from the British Library.

ISBN 978-0-2414-6777-0

Printed and bound in China

For the curious

www.dk.com
www.LEGO.com

MIX
Paper from
responsible sources
FSC™ C018179
www.fsc.org

This book was made with
Forest Stewardship Council™
certified paper – one small
step in DK's commitment to
a sustainable future.